APPLIED
GENETICS
in Healthcare

A handbook for specialist practitioners

D1404101

APPLIED GENETICS
in Healthcare

A handbook for specialist practitioners

Heather Skirton

PhD MSc Dip Counselling
RGN RM Registered Genetic Counsellor

Reader in Health Genetics, Faculty of Health and Social Work,
University of Plymouth, Plymouth, UK

Christine Patch

PhD MA(ed) SRN

Senior Research Fellow, Public Health Sciences, School of Medicine,
University of Southampton
Genetic Nurse Counsellor, Wessex Clinical Genetics Service

Janet Williams

PhD MA
RN FAAN

Kelting Professor of Nursing, The University of Iowa, Iowa City, Iowa, USA

Published by:

Taylor & Francis Group

In US: 270 Madison Avenue
           New York, NY 10016

In UK: 4 Park Square, Milton Park
           Abingdon, OX14 4RN

© 2005 by Taylor & Francis Group

ISBN: 1-8599-6274-2

Library of Congress Cataloging-in-Publication Data

Skirton, Heather.
  Applied genetics in healthcare : a handbook for specialist practitioners / Heather Skirton, Christine Patch, Janet Williams.
    p. ; cm.
  Includes bibliographical references.
  ISBN 1-85996-274-2 (alk. paper)
  1. Medical genetics--Handbooks, manuals, etc. 2. Human genetics--Handbooks, manuals, etc.
    [DNLM: 1. Genetics, Medical. 2. Genetic Counseling. QZ 50 S638a 2005]
  I. Patch, Christine. II. Williams, Janet K. III. Title.
  RB155.S525 2005
  616'.042--dc22
                              2005009369

Editor: Elizabeth Owen
Editorial Assistant: Chris Dixon
Senior Production Editor: Georgia Bushell
Typeset by: Phoenix Photosetting, Chatham, Kent, UK
Printed by: Cromwell Press, Trowbridge, Wiltshire, UK

Printed on acid-free paper

10 9 8 7 6 5 4 3 2 1

Taylor & Francis Group
is the Academic Division of T&F Informa plc.

Visit our web site at http://www.garlandscience.com

Contents

3 Working practically in genetic healthcare

4 Working professionally in genetic healthcare

5   Working to support families

10 X-linked inheritance

11 Familial cancer

12 Chromosomal and non-traditional patterns of inheritance

Preface

Increasingly, genomics is having an impact on mainstream healthcare. All health professionals will therefore be required to understand basic genetic concepts, but the depth of knowledge required will vary according to the role of the practitioner and the setting in which he or she works. The genetic knowledge and skills required to practice could be differentiated into three levels.

Foundation level

At this level, the practitioner needs to understand basic genetic principles, be able to detect situations in which there may be a genetic influence on the disease state and understand how to obtain specialist advice and information. This level of practitioner would include many general nurses and allied health professionals.

Advanced level

Practitioners in this group are generally specialists in a particular field of healthcare that requires knowledge of genetics in specific topics. In addition to understanding basic genetic principles, a practitioner at this level requires an in-depth knowledge of the genetic influence on the types of conditions involved in their realm of expertise. This group would include fetal medicine midwives or obstetric nurses, pediatric nurses and oncology nurses. It might also include some allied health professionals, such as speech and language therapists who work with patients with neurodegenerative conditions. A list of the types of work and knowledge required for some of the practitioners in this group is included in Appendix I.

Specialist genetics level

General specialist practitioners working in the field of genetics, such as genetics nurses or genetic counselors, would be included in this category.

Using this book

Skirton and Patch's first book, *Genetics for Healthcare Professionals*, was aimed at meeting the needs of practitioners at foundation level, but could be used as a first text for any healthcare practitioner who is new to genetics. This book includes information at a more advanced level and is chiefly written as a handbook for those training as genetics specialists. However, for completeness, some basic genetic science is also included in Chapter 2, with relevant sections expanded in later chapters where they apply. Many sections of the book will be relevant to advanced practitioners in other specialist areas of healthcare and the authors suggest that those readers select from the book those sections that will be relevant to their specific areas of expertise.

A glossary of terms is provided at the end of the book and the list of website resources is comprehensive. It is hoped that experienced practitioners will also use the book as a resource handbook and as a source of references for wider reading on specific topics of interest.

Abbreviations

Abbreviation	Definition
ABGC	American Board of Genetic Counseling
AFP	alphafetoprotein
AGNC	Association of Genetic Nurses and Counsellors
APNG	Advanced Practice Nurse in Genetics
APOE	apolipoprotein
AS	Angelman syndrome
ASHG	American Society for Human Genetics
AT	ataxia telangectasia
CADASIL	cerebral autosomal dominant arteriopathy with subcortical infarcts and leukoencephalopathy
CF	cystic fibrosis
CFTR	cystic fibrosis transmembrane receptor
CPEO	chronic progressive external ophthalmoplegia
DMD	Duchenne muscular dystrophy
DNA	deoxyribonucleic acid
DRPLA	dentatorubral-pallidoluysian atrophy
FAP	familial adenomatous polyposis
FISH	fluorescent *in situ* hybridization
G6PD	glucose-6-phosphate dehyrogenase
GCN	Genetics Clinical Nurse
GNCC	The Genetic Nursing Credentialing Commission
HBOC	hereditary breast and ovarian cancer
hCG	human chorionic gonadotrophin
HD	Huntington disease
HIF	hypoxia-inducible factor
HIPPA	The Health Insurance Portability and Accountability Act
HLA	human leukocyte antigen

HNPCC	hereditary non-polyposis colon cancer
HRT	hormone replacement therapy
LDLR	Low-density lipoprotein receptor
JCMG	Joint Committee on Medical Genetics
LDLR	Low-density lipoprotein receptor
LHON	Leber hereditary optic neuropathy
LS	Leigh syndrome
MCADD	medium-chain acetyl-CoA dehydrogenase deficiency
MELAS	mitochondrial encephalomyopathy with lactic acidosis and stroke-like episodes
MERRF	myoclonic epilepsy with ragged-red fibers
MMR	mismatch repair
mRNA	messenger RNA (ribonucleic acid)
MSI	microsatellite instability
MTHFR	methylenetetrahydrofolate reductase
NARP	neurogenic weakness with ataxia and retinitis pigmentosa
NCHPEG	National Coalition for Health Professional Education in Genetics
NF	neurofibromatosis
NF1	neurofibromatosis type I
NF2	neurofibromatosis type II
NIH	National Institutes for Health
NSGC	National Society of Genetic Counselors
NT	nuchal translucency
OMIM	Online Mendelian Inheritance in Man
PAH	phenylalanine hydroxylase
PAPP-A	pregnancy associated plasma protein A
PCR	polymerase chain reaction
PFGE	pulse-field gel electrophoresis
PKU	phenylketonuria
PNF	plexiform neurofibroma

PSA	prostate-specific antigen
PWS	Prader-Willi syndrome
RFLP	restriction fragment-length polymorphism
RP	retinitis pigmentosa
SCD	sickle-cell disease
SMA	spinal muscular atrophy
SNP	single nucleotide polymorphism
TA	transactional analysis
uT3	unconjugated oestriol
UPD	uniparental disomy
VEGF	vascular endothelial growth factor
VHL	von Hippel Lindau syndrome
VNTR	variable number tandem repeat
XLRP	X-linked retinitis pigmentosa

1 Introduction to genetic healthcare

1.1 Defining genetic healthcare

This book is written as a handbook for those who provide genetic healthcare. Genetic healthcare is defined by the authors as any intervention by a health professional that is aimed at addressing the physical, psychological, cognitive, emotional, or social needs of an individual or family, in cases in which those needs are connected with the presence or risk of a genetic condition. These interventions are usually delivered by professionals with a nursing, genetic counseling, or medical training. Although the professional background will undoubtedly have some influence on the delivery of care, many of the competencies and skills that are required to provide such care are common to all groups of specialist practitioners.

Genetics became the focus of a specialist healthcare service – requiring a general genetics knowledge at an advanced level – in both the United Kingdom (UK) and the United States of America (USA) after World War II. However, before that time genetic healthcare was offered as part of the medical care open to families and individuals by doctors operating in other specialities. This was particularly evident in pediatrics, in which the genetic basis of the inborn errors of metabolism were identified about 100 years ago (Garrod, 1908), and in neurology, in which patients affected by adult-onset inherited conditions (such as 'Huntington's chorea' and Charcot Marie Tooth disease) were treated. Currently, genetic services are mainly provided through specialist genetic centers, or in disease-focused programs, and by professionals with specific appropriate training working with colleagues in closely related fields.

Delivery of appropriate genetic healthcare is based on a belief that individuals have an inherent right to be properly informed about the genetic risks and reproductive options that might affect them, and that they should be supported during any decision-making process (Clarke, 1997).

In many specialist genetic centers, a multi-professional team approach enables a variety of expertise to be utilized to enhance client care. This book is written as a handbook for practitioners who provide genetic services, whether in a genetic center or as part of a team that provides genetic healthcare to a group of patients with a specific genetic health need.

1.2 Definition of genetic counseling

The aim of genetic services is to assist people who are at risk of developing or carrying a genetic disease to live and reproduce as normally as possible (Pembrey, 1996). This involves making accurate diagnoses, a discussion of appropriate options for testing or reproduction, and offering psychosocial support to families using the service (Clarke, 1997).

The definition of genetic counseling that is accepted in North America and in the UK, as well as in many other countries, was written by the American Society for Human Genetics (ASHG) in 1975. Although more than 30 years have passed since it was first written, it still accurately reflects the extent of services provided under the title 'genetic counseling'. It states that genetic counseling is:

> 'a communication process which deals with human problems associated with the occurrence, or the risk of occurrence, of a genetic disorder in a family. This process involves an attempt by one or more appropriately trained persons to help the individual or family to [1] comprehend the medical facts, including the diagnosis, probable course of the disorder, and the available management; [2] appreciate the way heredity contributes to the disorder, and the risk of recurrence in specified relatives; [3] understand the alternatives for dealing with the risk of recurrence; [4] choose the course of action which seems to them appropriate in view of their risk, their family goals and their ethical and religious standards, and to act in accordance with that decision; and [5] make the best possible adjustment to the disorder in an affected family member and/or the risk of recurrence of that disorder' (American Society of Human Genetics, 1975).

1.3 Genetic services within a healthcare system

The organization of genetic services necessarily varies from country to country. It is affected by the structure of the healthcare system, funding of healthcare, routes for professional education, statutory regulation of healthcare professionals, and healthcare culture and traditions. In many countries, other healthcare issues dictate to a large degree – to the extent that genetics is, of necessity, a low priority at present. In this chapter, the professional practice of non-medically trained practitioners, such as genetics nurses and genetic counselors in the UK and USA, are described, but these are by no means the only models of practice.

Genetic services in Europe

Clinical genetic services are established in at least 29 different countries in Europe, although the level of service is extremely varied. In some countries, such as the UK, Belgium, Denmark, the Netherlands, and Norway, a team approach is taken and both medical and non-medical personnel are part of the team who provide genetic healthcare. In the UK and the Netherlands, most genetic counselors initially trained in nursing or a similar paramedical field, but in Belgium genetic counselors frequently have a background in psychology. Recently, a Master's degree course in genetic counseling was approved in France to train genetic practitioners in that country. There has been a powerful medical influence on the practice of genetics nurses and counselors within Europe, as doctors usually lead clinical genetic services. Whereas in the UK genetics is designated a medical specialty, this is not the case in other areas of Europe, and therefore genetic counseling may be undertaken by doctors trained in other related specialties, such as obstetrics or pediatrics.

Specialist genetic services are provided in the UK by teams working in Regional Genetics Centres that are publicly funded by the National Health Service. This means that clients are able to access services free of charge. The genetics team usually includes medically trained clinical geneticists and non-medical genetic counselors. In the UK, most genetic counselors have a background in nursing, although there are an increasing number who enter the field from non-clinical backgrounds after completing a postgraduate Master's degree in genetic counseling. All use the term 'genetic counselor' as a professional title.

Specialist practitioners are now registered as genetic counselors by the Association of Genetic Nurses and Counsellors (AGNC) Registration Board (Skirton *et al.*, 2003). Genetics nurses and counselors practicing in the UK are bound by the AGNC Code of Ethics and are eligible for registration if they are either a registered nurse with a Bachelor's or higher degree or have completed an approved Master's degree. The practitioner demonstrates competence by submission of a portfolio of evidence after at least 2 years of mentored experience in genetic healthcare (AGNC, 2004). The AGNC competencies are listed in Box 1.1.

Box 1.1 AGNC competencies for UK genetic counselors

Core competencies for the practice of genetic counseling

The genetic counselor is able to:

Communication skills
- Establish a relationship with the client and elicit the client's concerns and expectations.
- Elicit and interpret appropriate medical, family, and psychosocial history.
- Convey clinical and genetic information to clients appropriate to their individual clinical needs.
- Explain the options available to the client, including risks, benefits and limitations.
- Document information, including case notes and correspondence in an appropriate manner.
- Plan, organize, and deliver professional and public education.

Interpersonal, counseling and psychosocial skills
- Identify and respond to emerging issues of a client or family.
- Acknowledge the implications of individual and family experiences, beliefs, values, and culture for the genetic counseling process.
- Make a psychosocial assessment of the client's needs and resources, and provide support, ensuring referral to other agencies as appropriate.
- Use a range of counseling skills to facilitate the client's adjustment and decision-making.
- Establish effective working relationships to function within a multi-disciplinary genetics team and as part of the wider healthcare and social care network.

Ethical practice
- Recognize and maintain professional boundaries.
- Demonstrate reflective skills within the counseling context, and in personal awareness for the safety of clients and families, by participation in counseling/clinical supervision.
- Practice in accordance with the AGNC Code of Ethical Conduct.
- Present opportunities for clients to participate in research projects in a manner that facilitates informed choice.
- Recognize his or her own limitations in knowledge and capabilities and discuss with colleagues or refer clients when necessary.
- Demonstrate continuing professional development as an individual practitioner and for the development of the profession.
- Contribute to the development and organization of genetic services.

Critical thinking skills
- Make appropriate and accurate genetic risk assessment.
- Identify, synthesize, organize, and summarize relevant medical and genetic information for use in genetic counseling.
- Demonstrate the ability to organize and prioritize a caseload.
- Identify and support clients' access to local, regional and national resources and services.
- Develop the necessary skills to critically analyze research findings to inform practice development.

AGNC (2004) www.agnc.co.uk/Registration/competencies

Genetic services in the USA

As in the UK, genetic services are provided by a variety of healthcare professionals in the USA. However, in the USA, genetic healthcare services are provided by employers whose healthcare programs may be established on either a for-profit or not-for-profit status. These employers include academic medical centers, government-operated healthcare services, privately owned genetics laboratories, and privately owned healthcare practices. Funding of services is dependent on a variety of factors that are related to the client's financial and insurance status, and also on the fiscal basis of the service. Commonly, a person is referred for genetic services by his or her healthcare provider, but in some locations clients may self-refer.

A client who wishes to have a genetic assessment or genetic counseling from a specialist may receive these services in tertiary-care centers (such as major medical centers) that are owned either by academic institutions or private companies. One example is the care provided in a medical genetics clinic in a university-owned hospital, in which board-certified or credentialed genetic counselors, and also advanced practice genetics nurses, undertake practice. Genetic services are also provided in healthcare practices that focus on one aspect of health, such as cancer or prenatal care, for those whose need for service lies within that specific category (e.g. oncology services or reproductive healthcare).

Professional organizations define guidelines for practice as a genetics specialist and maintain responsibility for administering credentialing programs for genetics specialists. Registered nurses, genetic counselors and medical doctors can qualify to become certified or to be credentialed as genetics specialists in their own profession. Some states in the USA have, or are considering passing legislation to institute, state licensing rules and regulations for genetic counselors.

The professional role of the genetics nurse and the genetic counselor

Genetics nurses and genetic counselors share a common mission of providing genetic healthcare to individuals, families and communities. These roles encompass individual client healthcare, advocacy for individuals and their families, membership in multi-disciplinary healthcare systems, participation in program management and evaluation, and involvement in clinical genetics-focused research. Although the scope of genetic healthcare roles can be quite broad, documents that describe the roles of genetics nurses and genetic counselors specifically describe the clinical aspects of genetics nursing or genetic counseling practice. In many countries, such as the UK, the work is undertaken in a genetic center by persons of either a genetics nursing or genetic counseling background. However, in some areas of the USA, there is greater differentiation in the roles.

In the USA, the professional role of the genetics nurse is defined by the scope of practice of genetics nursing (ISONG, 1998) and the regulations defining nursing practice in state professional licensure rules and regulations. The professional role of the genetic counselor is based on the counselor's education, certification as a genetic counselor, and, to some degree, expectations in the genetic counselors' employment setting. In those states that have genetic counseling licensure rules and regulations, the scope of practice is defined by the licensure rules and regulations.

In 1998, The Genetic Nursing Credentialing Commission (GNCC) was established by members of the International Society of Nurses in Genetics for the purpose of establishing and providing credentialing of genetics nurses (Cook *et al.*, 2003). The GNCC developed criteria for credentialing of nurses at either the Genetics Clinical Nurse (GCN) or the Advanced Practice Nurse in Genetics (APNG) level. Each credential is awarded on successful completion of review of a portfolio that contains evidence of completion of requirements in education and practice in the nurse's area of genetic expertise.

Genetics nurses are licensed professional nurses that have had special education and training in genetics. The professional role of the genetics nurse has its foundation in nursing, and is based on theories of nursing, genetic biology, behavioral and medical sciences (ISONG, 1998). Application of this knowledge throughout the processes of assessment, identification of outcomes of care, interventions and evaluation is the responsibility of the genetics nurse. Specific aspects of genetics nursing practice are defined in the GNC and APNG credentialing process (Box 1.2). Nurses may participate in the administration and monitoring of therapeutics as specified by their professional nursing license (Greco and Mahon, 2003; ISONG, 1998).

The National Society of Genetic Counselors (NSGC) defines a genetic counselor as a healthcare professional who has a specialized graduate degree and experience in the areas of medical genetics and counseling (Box 1.3). Genetic counselors enter the field from a variety of disciplines, including biology, genetics, nursing, psychology, public health, and social work, and

Box 1.2 Definition of genetics nursing practice in the USA

Aspects of genetics nursing practice defined in the GNC and APNG credentialing process include:

- collecting and interpreting comprehensive client information, including a genetics family history;
- deriving a diagnosis based on assessment data;
- identifying client-sensitive outcomes;
- identifying genetic risk factors;
- providing client-centered teaching;
- coordinating healthcare services;
- promoting health for client, family, and community;
- using therapeutic communication skills to foster the client's and family's understanding about the genetic condition;
- providing information regarding management of health risks;
- engaging in systematized ongoing evaluation of client and family.

must graduate from an accredited master's level program in order to qualify for certification. Credentialing is achieved through successful completion of a credentialing examination, administered by the American Board of Genetic Counseling (ABGC). Beginning with certificates issued in 1996, certification of genetic counselors was limited to a 10-year period and a recertification program has been developed (American Board of Genetic Counseling, 2003).

Box 1.3 NSGC definition of the role of a genetic counselor

The NSGC definition states that genetic counselors work as members of a healthcare team with families in which a child has been born with a birth defect or whose members are at risk of a genetic condition, to:

- provide information and support to families;
- identify families at risk;
- investigate the problem present in the family;
- interpret information about the disorder;
- analyze inheritance patterns and risks of recurrence;
- review available options with the family;
- provide supportive counseling to families;
- serve as patient advocates;
- refer individuals and families to community or state support services;
- serve as educators and resource people for other healthcare professionals and for the general public.

It is not within the remit of this book to discuss the practice of clinical genetics in every country. Professional roles in the UK and USA are used as examples of practice. Practitioners in other countries will benefit from contacting the relevant professional associations in their own country or region. Table 1.1 details the larger professional societies and their websites. A comprehensive list can be found at the website of the Karolinska Insitutet (listed in 'Further resources' at the end of this chapter).

Table 1.1 Professional organizations and websites

Organization	Country or region	Website
Canadian Association of Genetic Counselors	Canada	
American Society for Human Genetics	USA	http://genetics.faseb.org/genetics/ashg/ashgmenu.htm
Association of Genetic Nurses and Counsellors	UK	http://www.agnc.co.uk
Australian Society of Genetic Counsellors	Australia, New Zealand	No website listed
British Society for Human Genetics	UK	http://www.bshg.org.uk/
Canadian Association of Genetic Counselors	Canada	http://www.cagc-accg.ca/
Danish Society of Medical Genetics	Denmark	http://www.dsmg.suite.dk/
European Society for Human Genetics	All European countries	http://www.eshg.org/
Human Genetics Society of Australasia	Australia, New Zealand	http://www.hgsa.com.au/
International Society of Nurses in Genetics	Worldwide membership	http://www.isong.org/
Italian Society of Human Genetics	Italy	http://sigu.univr.it/
Japan Society of Human Genetics	Japan	
Mexican Association of Human Genetics	Mexico	http://www.kumc.edu/gec/prof/ibero.html
National Society for Genetic Counselors	USA	http://www.nsgc.org
Netherlands Society of Human Genetics	The Netherlands	http://www.nav-vkgn.nl

1.4 Basic areas of genetics knowledge required by practitioners working in mainstream healthcare

In most healthcare systems, there is an opportunity for individuals to obtain some genetics information from his/her primary healthcare provider. This could include the person's nurse practitioner, nurse midwife or family physician who reviews the person's family history. All of these individuals might provide basic information about genetic testing, and/or making a referral to a genetics specialist. Guidelines for primary care providers, as well as healthcare professionals who are not genetics specialists, are provided by the National Coalition for Health Professional Education in Genetics (NCHPEG, 2001).

A basic knowledge of genetics at Foundation level is required by all healthcare practitioners. Work on the core competencies in genetics for nurses has defined the extent of knowledge that is required (NCHPEG, 2001; Yeomans and Kirk, 2004). As a minimum requirement, all healthcare professionals should have knowledge of:

- the common mechanisms and patterns of inheritance, related to both chromosomes and single genes;
- the means of access to the local genetic service for information and/or referral;
- ethical issues related to confidentiality and autonomous decision making;
- the complexity of psychosocial reactions to a genetic disorder, including cultural and religious issues;
- the realities of genetic testing, potential benefits, and limitations.

General nurses working in primary care provide a good illustration of the range of genetic information that may be utilized within the context of everyday practice. These nurses can fulfill a number of varied roles, including the roles of health visitor, practice nurse, district nurse and school nurse. Because of the nature of primary care, all sections of the community will be represented in the client group, but those with particular needs for healthcare will include parents and young children, the elderly, patients with chronic conditions and school-age children. For example, nurses working in the primary care setting will require genetics education when:

- contributing to the development of policy guidelines for screening for genetic conditions within the local community (e.g. familial bowel or breast cancer);
- caring for a child with developmental delay;
- advising clients on the relevance of family history (such as cancer, heart disease, or diabetes);
- offering pre-conceptional care to couples.

1.5 Genetic services provided by other specialists

Within all healthcare services, there are a number of professionals from different disciplines whose work requires an extensive knowledge of genetics as applied to a particular clinical field or client group. These practitioners need to be genetically literate to provide basic genetic information about inheritance patterns and potential risks for clients. However, an understanding of genetics is even more important in a healthcare climate in which molecular genetic testing enables accurate diagnoses to be made, and in which gene-based treatments, such as enzyme replacement therapy, are now being used.

It is also essential that specialists in genetics understand the roles of those working in other healthcare settings, to enable them to liaise effectively with colleagues for the client's benefit. In addition, the genetics specialist nurse or counselor is often required to contribute formally or informally to the education of others, and an understanding of the purpose of such education is required to deliver relevant information at an appropriate level. Practice guidelines for non-genetics specialists regarding genetics have been developed by professional organizations such as the Association of Women's Health, Obstetric and Neonatal Nursing (2001) and the American Academy of Pediatrics (1993–2004).

In Appendix I, the application of genetics by healthcare professionals who need a working knowledge of genetics to provide care for their particular patient group is described briefly. Clinical examples and the competencies for each group are included. The specialists described are used to indicate the requirement for sound genetics knowledge outside the realm of specialist genetics, and to give trainees in specialist genetics an insight into some of the contexts in which genetics knowledge may be applied in mainstream healthcare services.

1.6 Conclusion

In this chapter, the definition of genetic healthcare and the types of professionals who provide that care have been discussed. Specialist providers may have a generic practice or may focus on a particular client or disease group. Although originating from a variety of professional backgrounds and healthcare structures, specialists who provide genetic services have a number of common aims. In the following chapters, the fundamental principles of specialist practice will be described.

References

AGNC (2004) Competencies for UK genetic counselors. [Accessed March 4, 2004] www.agnc.co.uk/Registration/competencies

American Board of Genetic Counseling (2003) Recertification guidelines. [Accessed August 2, 2004] http://abgc.net/genetics/abgc/abgcmenu.shtml

American Society of Human Genetics Ad Hoc Committee on Genetic Counseling (1975) Genetic counseling. *Am. J. Human Genet.* **27**: 240–242.

Association of Women's Health, Obstetric and Neonatal Nurses (2001) Role of the registered nurse in support of patients as related to genetic testing. [Accessed August 2, 2004] http://www.awhonn.org

American Academy of Pediatrics (1993–2004). Committee on genetics. [Accessed August 2, 2004] http://aap.org

Clarke, A.J. (1997), Outcomes and process in genetic counselling. In: *Genetics, Society and Clinical Practice, First Edition.* (eds P.S. Harper and A.J. Clarke). BIOS Scientific Publishers, Oxford, Chapter 12.

Cook, S.S., Kase, R., Middelton, L. and Monsen, R.B. (2003) Portfolio evaluation for professional competence: Credentialing in genetics for nurses. *J. Profess. Nurs.* **19**: 85–90.

Garrod, A.E. (1908) The Croonian lectures on inborn errors of metabolism, lecture II: alkaptonuria. Lancet, **2**: 73–79.

Greco, K. and Mahon, S.M. (2003) Genetics nursing practice enters a new era with credentialing. *Internet J. Adv. Nurs. Pract.* **5**: 1523–6064.

ISONG, International Society of Nurses in Genetics, Inc. (1998) *Statement on the Scope and Standards of Genetics Clinical Nursing Practice.* American Nurses Association, Washington, DC, USA.

NCHPEG. National Coalition for Health Professional Education in Genetics (2001) Core competencies in genetics essential for all health-care professionals. [Accessed January 19, 2005] www.nchpeg.org.

Pembrey, M. (1996) *The Progress Guide to Genetics.* The Progress Educational Trust, London.

Skirton, H., Kerzin-Storrar, L., Patch, C., Barnes, C., Guilbert, P., Dolling, C., Kershaw, A., Baines, E. and Stirling, D. (2003) Genetic counsellors: a registration system to assure competence in practice in the United Kingdom. Comm. Genet. **6**: 182–183.

Yeomans, A. and Kirk, M. (2004) Genetics for beginners. *Nurs. Stand.* **18**: 14–17.

Further resources

The most helpful sources of information about the professional roles of genetic healthcare specialists are available through the websites of the national and international professional bodies. A comprehensive list can be found at the website of the Karolinska Institute.

http://www.kumc.edu/gec/prof/soclist.html

2 Foundations in genetic science

2.1 Introduction

The information provided in this chapter is intended to provide a scientific basis for the material in other chapters. For some practitioners, this information may be unfamiliar. Many practitioners will have studied genetics in depth, but may not previously have applied that knowledge in a healthcare setting. It is obviously not possible in this text to cover all the scientific knowledge that is required for competent practice in genetic healthcare, and appropriate texts for more detailed reading are included at the end of the chapter. The face of genomic healthcare is changing so rapidly that current knowledge on the genetic basis of particular diseases is usually best obtained via scientific websites hosted by reputable institutions, such as Online Mendelian Inheritance in Man (OMIM) or the National Institutes of Health (NIH). These sources can be updated almost instantaneously as new findings emerge. A number of these websites are listed at the end of the book.

2.2 The structure of the gene

During the nineteenth century, Mendel spoke of 'factors' that he believed to be responsible for the inheritance of physical characteristics. Now known as genes, these factors were found by Watson and Crick (1953) to exist within the double-helical structure of deoxyribonucleic acid (DNA). The human chromosome pattern was accurately described for the first time by Tijo and Levan (1956) and it is along the chromosomes that the genetic material is arranged in an organized sequence. Genes are composed of exons (coding sequences) that are separated by introns (non-coding sequences).

The DNA molecule is the basis of the human genome. The double-helix is formed by a backbone of sugars that are linked by paired bases. The pyrine bases (adenine (A) and guanine (G)) have two carbon-nitrogen rings, whereas the pyrimidine bases (cytosine (C), and thymine (T)) have only one such ring. The pairing of the bases is complementary – that is, cytosine will only bind to guanine, and thymine to adenine. This means that the sequence of bases along the single strand can be replicated by binding each base to its complementary base. Hydrogen bonds exist between the bases to stabilize the pairing.

Each monomer in the chain of polymers is called a nucleotide and comprises a sugar, a base, and a phosphate group. The order of the bases in one strand of DNA provides the template for single-stranded ribonucleic acid, which provides the code for the formation of the polypeptide or amino acid. Each amino acid is coded for by a sequence of three bases (a codon) (Table 2.1).

Note that thymine is replaced by uracil in RNA.

Table 2.1 Codons and corresponding polypeptide

				Abbreviation	Amino acid
U	UUU UUC } Phe UUA UUG } Leu	UCU UCC UCA UCG } Ser	UAU UAC } Tyr UAA UAG } STOP	UGU UGC } Cys UGA } STOP UGG } Trp	Ala Arg Asn Asp	Alanine Arginine Asparagine Aspartic acid
C	CUU CUC CUA CUG } Leu	CUU CCC CCA CCG } Pro	CAU CAC } His CAA CAG } Glp	CGU CGC CGA CGG } Arg	Cys Gln Glu Gly His Ile	Cysteine Glutamine Glutamic acid Glycine Histidine Isoleucine
A	AUU AUC AUA } Ile AUG Met	ACU ACC ACA ACG } Thr	AAU AAC } Asn AAA AAG } Lys	AGU AGC } Ser AGA AGG } Arg	Leu Lys Met Phe Pro Ser	Leucine Lysine Methionine Phenylalanine Proline Serine
G	GUU GUC GUA GUG } Val	GCU GCC GCA GCG } Ala	GAU GAC } Asp GAA GAG } Glu	GGU GGC GGA GGG } Gly	Thr Trp Tyr Val	Threonine Tryptophan Tyrosine Valine

2.3 The chromosomes

Mitosis and meiosis

The majority of genetic material is stored on chromosomes within the nucleus of the cell (the remainder being in the mitochondria). Apart from the gametes, human cells are diploid, having two copies of each of the autosomes (chromosomes 1–22), and two sex chromosomes. Mitosis is the term used to describe the production of diploid cells through the process known as the cell cycle (Figure 2.1). During the normal cell cycle, each chromosome is replicated, producing identical sister chromatids that are joined at the centromere. Centrioles move to each side of the cell, and spindle fibers form between the centrioles. A pair of sister chromatids is arranged on each spindle fiber. When the chromatids separate, each chromosome is pulled to the opposite side of the cell from its sister. The nucleus then divides, producing two nuclei that each contain 46 chromosomes. The newly created cells are called daughter cells.

The germ cells (ova and sperm) are haploid, having only one copy of each chromosome. These gametes are produced in the ovary or testicle through the process termed meiosis (Figure 2.2). In the initial stage of meiosis, each chromosome is replicated, resulting in four copies of each chromosome. Cell division then takes place, resulting in two copies of each chromosome that are

Figure 2.1 Mitosis

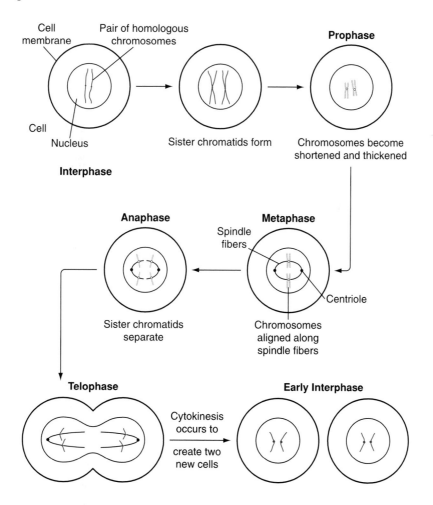

Figure 2.2 **Meiosis**

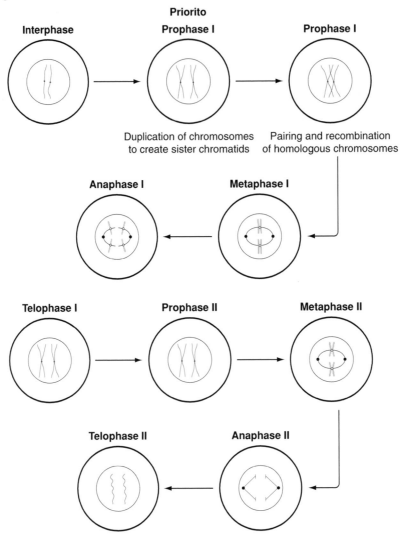

joined at the centromere. These sister chromatids line up along the spindle fibers and the cells split again, leaving only one copy of each chromosome in each of the gametes.

In humans and other animals, the products of meiosis differ between males and females. In the male, four spermatazoa will be produced. Meiosis in the female results in one ovum and two polar bodies.

From the perspective of genetic disease, disjunction of the two chromatids is a key process (Figure 2.3). Failure of chromatid disjunction – for example, non disjunction – can result in an abnormal number of chromosomes in the gamete (Figure 2.4). This is the precursor to aneuploidy in the embryo.

The clinical result of a non disjunction event is aneuploidy (Table 2.2), a situation in which the conceived embryo has an abnormal number of chromosomes. This is often incompatible with the development of the fetus, so full aneuploidy of some chromosomes (e.g. trisomy 1) is not seen in liveborn children. In other rare cases, children may survive with a mosaic chromosome structure that consists of two or more populations of cells, some of which have a normal karyotype and some of which contain an additional chromosome. Aneuploidies that could result in a liveborn child are listed in the table overleaf.

Figure 2.3 Disjunction

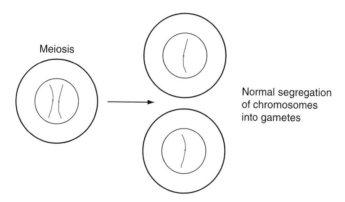

Normal segregation of chromosomes into gametes

Figure 2.4 Non disjunction

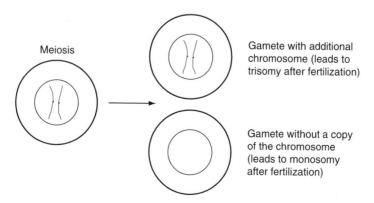

Gamete with additional chromosome (leads to trisomy after fertilization)

Gamete without a copy of the chromosome (leads to monosomy after fertilization)

Table 2.2 Aneuploidy – clinical examples

Chromosomal abnormality	Clinical examples
Trisomy	Trisomy 21 (Down syndrome) Trisomy 18 (Edward syndrome) Trisomy 13 (Patau syndrome) Children with full trisomy 18 and 13 are very unlikely to survive infancy.
Monosomy	Partial monosomy of autosomes may occur, but complete monosomy of any autosome is incompatible with life.
Additional sex chromosome	47, XXY (Klinefelter syndrome) 47, XXX (Triple-X syndrome)
Single sex chromosome	45, X (Turner syndrome)

Recombination

During the first meiotic division, each pair of chromatids is connected at junctions called chiasmata (singular 'chiasma'). The chromosome inherited by that individual from their mother is linked with the chromosome inherited from their father. There is some exchange of the chromosomal material between the homologous pairs, the resultant two chromatids being a combination of maternal and paternal chromosomes.

As the numbers of different combinations of the chromosomal material is almost infinite, each child's genotype will be a unique combination of the characteristics of all four grandparents. A disruption to the process of recombination can result in an abnormality of one or more of the chromosomes (Figure 2.5; Table 2.3). Changes in the balance of the chromosomal material will generally have serious implications for the intellectual and physical developmental of the individual. However, not all of these abnormalities will result in phenotypic changes. If the amount of chromosomal material remains balanced, a change in the location of the material may not have any effect on the health of the individual. It may, however, be of importance when the individual has children, as the haploid product of the cells may be unbalanced.

Imprinting

In general, both copies of an autosomal gene will be equally effective, regardless of whether the gene was inherited from the mother or father. However, there are specific areas of the chromosomes at which the maternally

Figure 2.5 Recombination

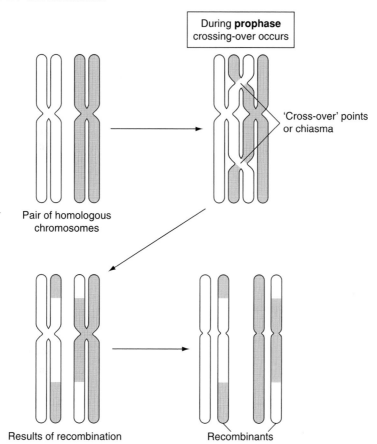

During **prophase** crossing-over occurs

Pair of homologous chromosomes

'Cross-over' points or chiasma

Results of recombination

Recombinants

Table 2.3 Chromosomal rearrangements and clinical examples

Rearrangement	Clinical example
Robertsonian translocation – results from the joining of two acrocentric chromosomes (Figure 2.6)	Robertsonian translocation of chromosomes 14 and 14, 14 and 15, or 14 and 21 are relatively common
Balanced reciprocal translocation – results from exchange of chromosomal material between any two chromosomes (Figure 2.7)	Can occur between any two chromosomes
Deletion of portion of chromosome	Microdeletion of 22q causes velocardiofacial syndrome
Duplication of portion of chromosome	Can occur in any chromosome
Inversion – paracentric or pericentric (Figures 2.8 and 2.9)	Can occur in any chromosome
Insertion of portion of chromosome (Figure 2.10)	Can occur in any chromosome

Figure 2.6 Robertsonian translocation

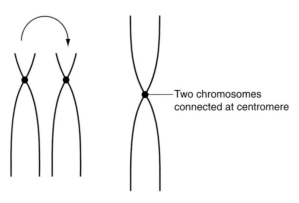

Two chromosomes connected at centromere

Figure 2.7 Reciprocal translocation

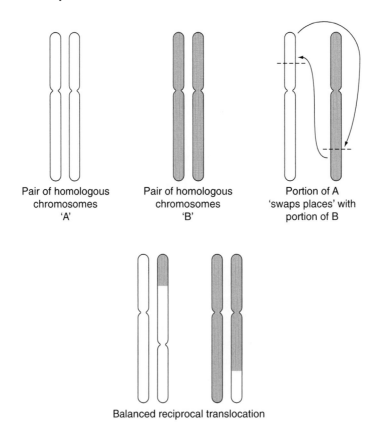

Pair of homologous chromosomes 'A'

Pair of homologous chromosomes 'B'

Portion of A 'swaps places' with portion of B

Balanced reciprocal translocation

and paternally derived copies of a gene are expressed differently. Imprinting is the term used to describe the situation in which a copy of the gene from the parent of a particular sex is 'switched off'. For example, the genes that are implicated in Prader-Willi syndrome (PWS) and Angelman syndrome (AS) are located in close proximity to each other on chromosome 15.

Figure 2.8 Paracentric Inversion

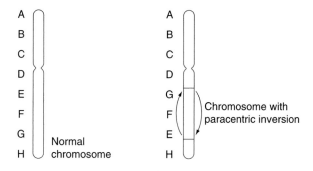

Figure 2.9 Pericentric Inversion

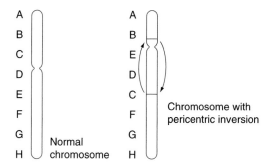

Figure 2.10 Insertion

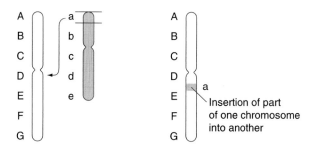

The maternal copy of the gene is imprinted in the PWS region and the paternal copy is imprinted in the AS region. Absence of a normal gene from the father, due to a deletion or uniparental disomy, will result in PWS. A child who does not inherit a normal copy of the gene from the mother will develop AS. This is discussed in more detail in Chapter 12. Many imprinted genes are involved in regulating fetal growth, and it is believed that paternally derived genes may be imprinted to protect the mother from the aggressive demands for nutrients from the fetus (Strachan and Read, 1999). Conversely, some maternal genes are imprinted to enable the fetus to grow adequately.

Uniparental disomy

Uniparental disomy (UPD) occurs if a child inherits two copies of a region of the chromosomal material from one parent. This may apply to a full or partial chromosome. If the UPD includes an imprinted region, then a genetic condition may result from the lack of genetic contribution from the relevant parent. UPD can follow a non disjunction event that results in the ovum or sperm having two copies of a particular chromosome. After fertilization, the blastocyst has three copies of that chromosome, but this imbalance can be corrected by the loss of one copy of that particular chromosome (termed trisomic rescue). However, if the copy that is 'lost' comes from the parent that contributed only one copy, the two remaining copies will have originated from the same parent. This concept is also discussed in Chapter 12.

2.4 Investigating the chromosome structure

Karyotyping

Karyotyping is the term used to describe the study, by a cytogeneticist, of the chromosome constitution in tissue derived from an individual or fetus. The chromosome structure is not easily seen during interphase, so cells are cultured in a tissue medium so that they can be studied during mitotic division. When sufficient numbers of cells are undergoing metaphase, the process is 'frozen' by adding an agent such as colchicine. This effectively destroys the spindle fibers and the next stage of mitosis (anaphase) is not reached. Without the supporting spindle fibers, the chromosomes spread more evenly around the cell nucleus (Figure 2.11). A salt solution added to the preparation swells the cells, further separating the chromosomes and enabling them to be viewed more easily. The chromosome preparation is 'fixed', placed on a slide, and viewed under the microscope.

Figure 2.11 Metaphase spread

Figure 2.12 G-banded karyotype

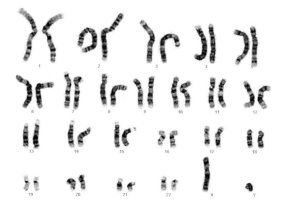

The use of staining techniques (such as G-banding) assists in the differentiation of each part of the chromosome, so that disruptions to the normal arrangement can be more easily detected (Figure 2.12).

When reporting the results of a karyotype, the cytogeneticist will always use a standard nomenclature devised by the International Standing Committee on Human Cytogenetic Nomenclature (1995). A cytogenetic report will include the total number of chromosomes observed, the type and number of sex chromosomes and any anomaly or abnormality of the chromosomes.

Fluorescent in situ hybridization

Fluorescent *in situ* hybridization (FISH) is a technique used in the laboratory to detect small deletions and rearrangements in a chromosome (Figure 2.13).

Figure 2.13 Microdeletions identified using FISH (fluorescent *in situ* hybridization)

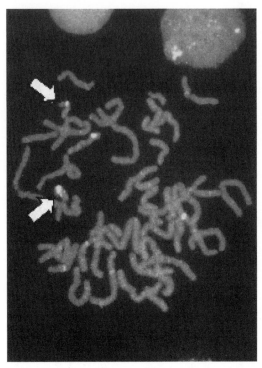

The technique can be used when the phenotype suggests a specific microdeletion or rearrangement. A probe that matches the normal DNA sequence on the portion of the chromosome to be studied is attached to a fluorescent marker. When the probe is mixed with the chromosome preparation, it adheres to the normal chromosome(s), sending a colored signal. If an autosomal deletion is present, the probe will fail to attach itself to the abnormal chromosome. Rearrangements can also be detected if the signal is not in the expected position on the chromosome.

Single gene inheritance

In a species, the single gene responsible for a trait occupies a specific locus on a chromosome. Each single copy of the gene at that locus is termed an allele. One allele is inherited from each parent. Within the human population, it is possible for there to be many different versions of a normal gene at one locus – these are termed polymorphisms. The differences between alleles can therefore be due to polymorphic variation or mutation. An alteration in the gene sequence is said to be a mutation if it occurs in less than 1% of the population, regardless of whether the change has a phenotypic effect. However, in genetic healthcare, the term mutation is often used in the context of a disease-causing alteration. If an organism carries two identical alleles at a locus, they are said to be homozygous, whereas if they carry two different alleles at a locus they are heterozygous.

In human genetics, there are five patterns of traditional inheritance. These are often referred to as Mendelian patterns after the father of genetic science, Gregor Mendel. These are:

- autosomal dominant;
- autosomal recessive;
- X-linked recessive;
- X-linked dominant;
- Y-linked.

These are described fully in future chapters (see Chapters 8–10) so will not be repeated here.

Expression and penetrance

The DNA sequence in the gene acts as a template for the production of messenger RNA (messenger ribonucleic acid, mRNA) (Figure 2.14). The mRNA leaves the nucleus and binds to ribosomes, where it directs the production of proteins. The string of amino acids is built up as each trinucleotide sequence in the mRNA codes for one amino acid. Expression of

Figure 2.14 Transcription

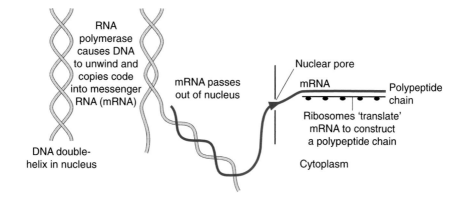

a gene is dependent on the stage of development of the organism and the function of the cell.

If a gene mutation always results in a phenotypic change, it is said to be fully penetrant. For example, the Huntington disease (HD) mutation (an expansion in the huntingtin gene) is fully penetrant; all those individuals who have the expansion will develop HD, provided they do not die of another cause before the disease develops. Conversely, lack of penetrance may mean that individuals who have inherited a mutation do not subsequently develop the disease. For example, approximately 20–30% of women who have a mutation in the BRCA1 gene do not develop breast cancer or ovarian cancer. Other differences in the way that a condition is manifest can be explained by variable expression. Variable expression refers to the phenomena whereby individuals may experience different signs and symptoms of the condition, which may also vary in severity. Autosomal dominant conditions, such as neurofibromatosis, tuberous sclerosis and Marfan syndrome, demonstrate this phenomenon. For example, some family members with neurofibromatosis may have only mild skin manifestations, such as café au lait patches, whereas other members of the same family may have serious effects, such as sarcoma. The concepts of penetrance and expressivity are discussed further in Chapter 8.

2.5 Mutations in the gene

Types of mutation

A mutation is an alteration in the normal sequence of DNA within the gene that may be inherited or occur sporadically. Mutations can be due to the effects of chemical or other physical damage (such as radiation) to the cell, but can also occur because of an error in DNA replication during mitosis or meiosis. However, once a mutation exists, it is replicated in all daughter cells that arise from the mutated cell.

Mutations may be harmless (in the sense that they do not cause disease), either because they occur in introns or because they do not alter the amino-acid sequence. The latter are sometimes called silent mutations and, because they do not increase morbidity, they tend to remain in the genotype as polymorphisms.

Alterations of a single base are called point mutations, whereas those that affect a larger portion of the gene sequence are called gross mutations.

The types of mutation are listed in Table 2.4 but discussed in greater detail in the relevant sections of later chapters that deal with specific inheritance patterns.

Missense mutations may have a serious phenotypic effect, but if only one amino acid is altered the effect may be minimal. Nonsense mutations cause the translation of the mRNA to end prematurely, resulting in a shortened protein. In most cases, this happens because the mutation causes the codon to be read as a STOP codon. Nonsense mutations usually have a serious effect on the encoded protein and may cause a mutant phenotype.

Insertion or deletion of one or more bases may also cause mutant phenotypes. If the alteration is a multiple of three, the reading frame of the DNA sequence may remain constant, although there will be extra or missing amino acids. If the reading frame is disrupted, the sequence of amino acids downstream from the mutation will read differently. Frameshift mutations usually have a serious effect on the resulting protein.

Table 2.4 Types of mutation

Mutation type	Example	Effect on amino acid
Point mutations – single base substitution		
Nonsense mutation – creates premature stop codon	TTG to TAG	Leu to STOP codon
Missense mutation – alters the encoded amino acid	GGC to AGC	Gly to Ser
Frameshift mutation – insertion or deletion of extra bases to alter the reading frame of the gene	CGG GTT TTG AAG to CGG GTT TTT GAA (insertion of T)	Arg Val Leu Lys to Arg Val Phe Glu
Gross mutations		
Expansion	CAG CAG TTA TTA AAG to CAG CAG CAG CAG CAG CAG TTA	Additional glutamines inserted in chain of amino acids
Deletion – from one base to entire gene or more	Entire exons may be deleted (e.g. in the dystrophin gene)	
Duplication – addition of material copied from a part of the DNA sequence	CGG TTA TAT GGC AGA ATT to CGG TTA TAT CGG TTA TAT GGC AGA ATT	Arg Leu Tyr Gly Arg Ile to Arg Leu Tyr Gly Leu Tyr Gly Arg Ile
Inversion – segment of DNA excised and re-inserted in opposite alignment	CGG TTA TAT GGC AGA ATT to CGG AGA GGC TAT TTA ATT	Arg Leu Tyr Gly Arg Ile to Arg Arg Gly Tyr Leu Ile

2.6 Mitochondrial inheritance

In addition to the DNA found in the nucleus of the cell, a relatively small amount of DNA exists in the mitochondria to control some mitochondrial activity. The mitochondrial DNA is inherited only from the mother, as the mitochondria in the spermatazoa is located in the tail, which is discarded at fertilization.

The number of mitochondria in human cells varies according to the type of cell. For example, neural cells have more mitochondria than skin cells, and therefore disorders caused by mitochondrial mutations often affect the neuromuscular system. The mitochondrial DNA is more prone to sporadic change than nuclear DNA and for this reason it is a mutation hotspot. The cell and its components, including the mitochondrion, are illustrated in Figure 2.15.

Additional information on mitochondrial inheritance is included in Chapter 12.

Figure 2.15 The cell, including the mitochondrion

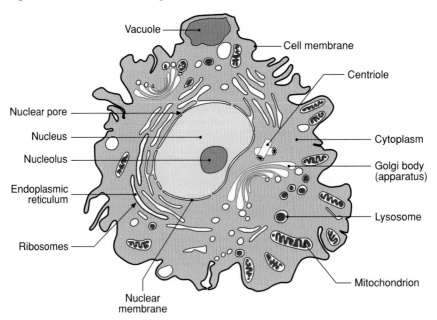

2.7 Polygenic inheritance

Some human characteristics are controlled by a single gene and a mutation in that gene produces a particular phenotype. Other traits are obviously influenced by a number of different genes, in conjunction with environmental factors. The latter usually follow the normal distribution and are described as polygenic or multifactorial. The role of genetic predisposition and environmental factors has been effectively demonstrated by studies of twins, particularly in cases in which the twins have been raised separately. When twins are 100% concordant with respect to a condition, the effect must be due to an inherited factor, probably a single gene, but when there is a level of discordance, the role of the environment must be questioned.

Whereas traits such as height and intelligence are continuous variables, other binary traits (that is, the person does or does not have the characteristic) may be subject to a threshold effect. The fetus or individual is susceptible to a condition because of a particular combination of genes within their genotype, but an external influence can trigger the change from normal to abnormal development or status. Examples of these types of conditions are coronary heart disease, non-insulin-dependent diabetes, schizophrenia, and neural tube defects (Reilly, 2004). Because the environmental element may be controlable, many multifactorial conditions may be susceptible to intervention, particularly if those individuals at high risk are identified by genetic or biochemical testing.

The role of genetic influence in common diseases is covered in depth in Chapter 13.

2.8 Laboratory testing

Specific laboratory testing techniques are described where relevant in later chapters. The most commonly used are listed in Table 2.5 for reference.

Different methods of testing may be used to identify disease genes. This topic is covered in Chapter 9, section 9.7.

2.9 Conclusion

This chapter has included the basic information that is needed by the practitioner in genetic healthcare. Although the basic concepts still remain, new advances in understanding the genome are leading to developments in diagnosis, therapy and classification of disease.

Table 2.5 Techniques for genetic laboratory testing

Name of test	Potential application	Examples of clinical use
Karyotype	To investigate number and gross structure of chromosomes	Pre-natal diagnosis of aneuploidy Identification of translocation carrier
Fluorescent *in situ* hydridization (FISH)	To investigate number of specific chromosomes To detect specific abnormalities of the chromosomes at micro level	Detection of specific aneuploidy (e.g. Trisomy 21) Detection of chromosomal micro-deletion (e.g. 22q11)
Southern blotting – using restriction fragment-length polymorphisms (RFLPs)	To examine the alleles inherited by an individual at a specific locus, to detect variation due to polymorphism or mutation Further information in section 4.6	Linkage studies to determine carrier or affected status of an individual Detection of gross mutation (e.g. expanded fragment or deletion)
Single nucleotide polymorphism (SNP)	To examine the nucleotide sequence of a segment of a gene	Detection of point mutations in a gene
Variable number tandem repeat (VNTR) polymorphisms.	Creates a 'genetic' profile	Identification of an individual's relationship within a family

Study Questions

1. Using the recognized nomenclature, give the four types of sex chromosome aneuploidy where the total number of chromosomes is either 45 or 47.
2. Using the recognized nomenclature, how would the chromosome structures of the following patients be reported by the cytogenetic laboratory?
   (a) Female with cytogenetically normal chromosome structure.
   (b) Male fetus with trisomy 8.
   (c) Female with a Robertsonian translocation involving chromosomes 13 and 14.
3. Name four types of genetic point mutations.
4. Select a disease or condition in which you are professionally interested. Using the resources desbribed at the end of the chapter (or others), identify the types of gene mutation that have been identified as pathogenic mutations in individuals affected by the condition.

References

International Standing Committee on Human Cytogenetic Nomenclature (1995) *International System for Human Cytogenetic Nomenclature* (ed. F. Mitelman). Karger, Basel.

Reilly, P.R. (2004) *Is It In Your Genes?* Cold Spring Harbor, New York.

Strachan, T. and Read, A.P. (1999) *Human Molecular Genetics,* Second Edition. BIOS Scientific Publishers, Oxford.

Tijo, H. and Levan, A. (1956) The chromosomes of man. *Hereditas* **42:** 1–6.

Watson, J.D. and Crick, F.H. (1953) Molecular structure of nucleic acids: a structure for deoxyribonucleic acid. *Nature* **171:** 737.

Further resources

Brown, T.A. (1999) *Genomes*. BIOS Scientific Publishers, Oxford.

Gardner, A., Howell, R.T. and Davies, T. (2000) *Human Genetics.* Arnold Publishers, London.

Guttman, B., Griffiths, A., Suzuki, D. and Cullis, T. (2002) *Genetics: A Beginner's Guide.* Oneworld Publications, Oxford.

Karolinska Insitutet [Accessed April 4, 2005] http://info.ki.se/index_en.html

Passarge, E. (1995) *Color Atlas of Genetics.* Thieme, New York.

Pritchard, D.J. and Korf, B.R. (2003) *Medical Genetics at a Glance.* Blackwell Science, Oxford.

Reilly P.R. (2004) *Is It In Your Genes?* New York, Cold Spring Harbour.

Strachan, T. and Read, A.P. (1999) *Human Molecular Genetics*, Second Edition. BIOS Scientific Publishers, Oxford.

Winter, P.C., Hickey, G.I. and Fletcher, H.L. (1998) *Instant Notes in Genetics.* BIOS Scientific Publishers, Oxford.

3 Working practically in genetic healthcare

3.1 Introduction

This chapter is aimed at enabling the health professional to develop the basic skills that are required to practice clinically in a genetics setting. As the basic tool of any genetic assessment is the family tree, this will be discussed in detail. Other clinical applications of genetic technology, such as prenatal and predictive testing, are described. Methods of risk calculation based on inheritance patterns, family history, and biochemical or molecular test results are also explained in this chapter.

3.2 Constructing a family tree

Background

The purpose of drawing a family tree, or pedigree, is to document the family medical history as accurately and clearly as possible. A considerable amount of information can be recorded in this graphic form, and the use of commonly recognized symbols enables other health professionals to correctly interpret the information (Figure 3.1).

However, although the main objective of drawing the family tree is to document biomedical information, the process inherently conveys information about the emotional, psychological and social status of the client and other family members. Telling stories about the health concerns or problems of family members may awaken painful memories, particularly if these involve loss or death. The social structure of the family will also be revealed through this process; death, separation, divorce, adoption, and miscarriage are just some of the sensitive and potentially painful issues that may arise in any family story. It is essential that the

Figure 3.1 **Accepted symbols for drawing the family tree**

Pedigree symbols

There is a generally accepted set of symbols that are used to convey information about the family structure via a pedigree. The word 'pedigree' derives from the French for 'foot of a crane', as the original pedigrees resembled the structure of a bird's foot.

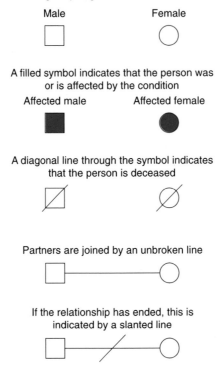

Male   Female

A filled symbol indicates that the person was or is affected by the condition

Affected male   Affected female

A diagonal line through the symbol indicates that the person is deceased

Partners are joined by an unbroken line

If the relationship has ended, this is indicated by a slanted line

practitioner who takes the family history is able to do so sensitively, using basic listening and counseling skills.

There is a trend in healthcare to use what is termed a genogram to display social as well as biological data about the family and relevant relationships (Hockley, 2000; Keiley *et al.*, 2002). However, it is the authors' view that genograms are not usually appropriate in a specialist genetics setting. The genogram will represent the view of family relationships from the perspective of the person who initially provides the information, but may conflict with the views of others within the family. As the pedigree will be used in the care of a number of family members, it is preferable that social information is recorded separately, where applicable. The pedigree is also often used over many years (with appropriate amendments), and as such should only contain

Figure 3.1 continued

If a child's parents have never been a couple,
the line between them is interrupted

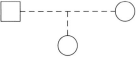

Consanguinity is marked by a double
line joining the two partners

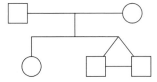

Twins are denoted

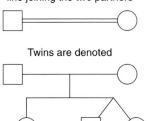

If twins are monozygotic (identical) a
horizontal line is drawn between them

A child who is stillborn is marked by a small circle, as is an
abortion (whether spontaneous or induced). The gestation
of the pregnancy is usually indicated if possible

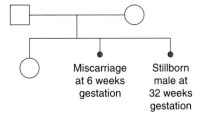

Miscarriage   Stillborn
at 6 weeks   male at
gestation   32 weeks
               gestation

verifiable and factual information rather than the impression or opinion of an
individual on the state of relationships in the family.

Ethical issues in taking a family tree

Confidentiality   Usually, a family medical history will be provided by the
consultand (person having the consultation), but there may be some other
members of the family present who contribute. However, information will

Figure 3.1 continued

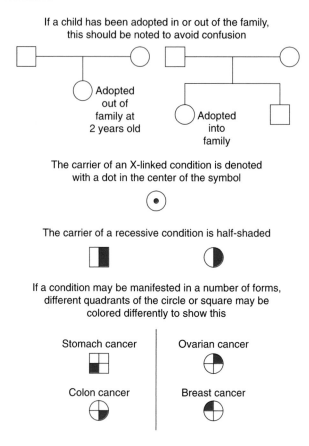

If a child has been adopted in or out of the family,
this should be noted to avoid confusion

Adopted
out of
family at
2 years old

Adopted
into
family

The carrier of an X-linked condition is denoted
with a dot in the center of the symbol

The carrier of a recessive condition is half-shaded

If a condition may be manifested in a number of forms,
different quadrants of the circle or square may be
colored differently to show this

Stomach cancer   Ovarian cancer

Colon cancer   Breast cancer

inevitably be disclosed about relatives who are not present and who may not have given their consent to disclosure. This information should not be shared (e.g. with other family members who may not be aware of it) without the consent of the person it concerns. Similarly, some of the information provided by the consultand may be highly confidential, and should not be shared with other family members unless consent is expressly given. The genetics practitioner's practice in the USA is also governed by The Health Insurance Portability and Accountability Act (HIPAA) of 1996 privacy regulations. The clinician has the duty of not disclosing medical information without the signed consent of the patient (US Department of Health and Human Services, 2004).

Unexpected information A family will usually be seeking genetic information about a specific condition that is regarded by them to be a potential risk. However, when taking the family history, the practitioner may become aware of information about another genetic condition that could be significant. It can be difficult to deal with this unexpected information. Obviously, the practitioner will be aware of the potential detriment to the family if the information is not acted on, but also wishes to avoid creating additional anxiety for the family.

It can be useful to apply ethical principles and to try to weigh the potential benefit of discussing the unexpected condition against any harm that may be done. Some questions that may help the practitioner to decide what action is appropriate are:

- Has the family indicated that they are concerned about the condition?
- Has the family asked questions about the condition?
- Is there likely to be any significant health risk to family members?
- Are any of those risks avoidable?
- Is there any treatment or screening available from which the family could benefit?

If the answer to any of these questions is in the affirmative, then undoubtedly the issue should be discussed with the family, but not necessarily during the current meeting. It is often helpful to have time for discussion with colleagues before raising the issue again with the family. Allowing time for the family to think about what has been raised can enable the persons concerned to start to process the information before deeper discussion occurs.

A practical format for drawing the family tree

It is common for individuals or families to be eager to facilitate the genetics counseling process by providing a flood of information about the medical history of the family. When taking a family tree, it is important that an organized process is followed to ensure that the recorded information is correct and complete, while maintaining flexibility to allow for the needs of each situation (Bennett, 1999). Each practitioner will develop his or her own style, but the following hints based on the authors' experience can be used as a guide.

1. Obtain details about the consultand first, followed by the members of his or her family (children, parents and siblings). It has been shown that clients have some anxiety about being able to respond to the practitioner's request for information (Skirton, 2001). It is therefore usually helpful to start with those who the client knows best to build confidence and establish the pattern of questioning. Of course, the first-

degree relatives are also likely to be of greatest importance from a genetic viewpoint.

2. Focus properly on one side of the family before moving to complete the other side. It is usually good practice to obtain details as far back as the client's grandparents. Information about relatives in even earlier generations may be useful, particularly if the condition is autosomal dominant or X-linked. However, when dealing with a recessive or sporadic condition, a three-generation tree is usually sufficient and most individuals will have very little information on family members who are older than their grandparents.

3. Specifically ask if there are relatives who have died, as clients do not always appreciate that deceased relatives are also of relevance to the genetic history. Similarly, ask about any pregnancy losses or neonatal deaths.

4. Clients may provide information about relatives who are socially rather than biologically related (such as step-siblings). It is important to include them on the pedigree for two reasons. First, to acknowledge and record the social structure of the family; and second, so that any future enquires made on behalf of the person result in the non-biological relationship being recorded. However, the type of relationship must be made clear on the pedigree. Children who are adoptees should be included, with a note indicating whether they have been adopted into or out of the family.

5. A question should always be asked about consanguinity in the family, as of course this would influence the potential for offspring being affected by a recessive condition. In many cultures, consanguineous marriage is the norm, but in others there may be a perceived stigma attached to cousin marriage and the question therefore needs to be phrased sensitively.

6. In the course of family history-taking, it is inevitable that private and sensitive information may be disclosed. This could include information about previous pregnancies, termination of pregnancies, paternity or relationships. Clients are often unsure whether this type of information is relevant and will offer it to the practitioner, but it is not always necessary to record it on the family tree. The practitioner should use his or her discretion as to whether the information is useful in determining a health risk to the consultand or other family members.

7. To ensure confidentiality between different branches of the family, a new family tree can be drawn when counseling people from each branch of the family. This enables the practitioner to determine what information is shared and therefore avoid breaching confidentiality.

8. After completing the tree, the client should be asked whether there is any information that he or she feels is important that has not been included. At each subsequent meeting, it is helpful to clarify whether any changes have occurred (e.g. births or deaths) so that the tree can be amended in accordance with this.

For further help with drawing the family tree, see the listed websites in 'Further resources' at the end of this chapter.

CASE STUDY – NATALIE

Natalie was referred to a genetics clinician for counseling following three miscarrriages that had occurred very early in pregnancy. She attended the genetics clinic alone because her husband Jim was serving in the armed forces and was away on duty. Natalie told the genetics nurse that she had a termination of pregnancy before meeting her husband, but that he was unaware of this. The genetics practitioner recorded this fact in the written notes (with a note about confidentiality), but did not record it on the pedigree as that would be used in subsequent sessions when Jim would be present. Karyotyping was performed on blood samples from Natalie and Jim. Natalie was found to carry a balanced translocation, which provided a probable explanation for her miscarriages.

CASE STUDY – GEORGE

George attended a clinic to discuss, with a genetic health professional, his history of Huntington disease (HD). When providing details of his family tree, he told the counselor that his 'father' and brother had HD. He also said that he had been told by an uncle that the man he had always regarded as his father may not have been his biological father. This was recorded on the pedigree as it may have been important for the interpretation of genetic test results. The counselor was also aware that he was unlikely to see others in the family as they resided in a different part of the country.

3.3 Risk calculation

Background

An integral part of the work of every genetic healthcare professional is to calculate the chance of occurrence of a particular condition in an individual or his or her offspring. In genetic healthcare settings, chance is generally referred to as recurrence risk, and it has been repeatedly shown that the numerical value is only one component of the information a person uses in their assessment of their personal risk. Other factors such as their personal experience of the condition and the way in which they perceive the burden of the condition are also very important. However, a numerical value provides a guide as to the probability of the condition occurring. Research indicates that individuals who are at risk consistently overrate this probability (van Dijk *et al.*, 2003) and, although an accurate assessment is essential, the person may still have difficulty in integrating the new information into his or her decision-making.

In the following section, the basic methods of calculating numerical risk are described.

Risk calculation using mode of inheritance

The recurrence risk can be accurately calculated for an individual when:

- there is certainty about the diagnosis;
- the mode of inheritance of the condition is understood;
- the person's biological relationship to affected family members is known.

It should be clarified for the client that the risk applies to each pregnancy. It is commonly believed that if the risk is, for example, 1 in 2 and the first child is affected, that the second will not be affected because the 1 chance out of 2 of having an affected child has already occurred. Use of physical teaching tools and visual images can help to convey the concept of chance. For example, flipping a coin repeatedly demonstrates the concept that chance has no memory, and a pack of four marked cards can be used to explain recessive inheritance risks. It is important to use terminology that is understood by the client. For some people, using 'odds' ratios is helpful because of their familiarity with betting systems. For others, phrasing the chance in percentages may make more sense. Although it is usual to speak about the chance of a person being affected, it is also helpful to reverse the odds and give the client an idea of the chance of them being unaffected. Putting the risk into perspective is also useful. For example, a woman who has a 20% lifetime risk of developing breast cancer needs to know that the population risk is 8–13%. This knowledge allows her to form a view about her relative risk compared to others in her community.

Autosomal dominant inheritance pattern   A mutation in an autosomal dominant gene will generally cause the person who has inherited the mutation to be affected by the condition. The affected person has two copies of the gene: one allele with a mutation and one that is normal. At meiosis, only one copy of each gene passes into the gamete from the parent; therefore, any offspring of an affected person has a 50% chance of inheriting the mutation. Due to variability of expression and reduced penetrance (see Chapter 8), a person who may not appear to have the condition may actually have inherited the mutation and his or her children will be at a 50% risk of developing the condition. Careful physical examination and, if possible, mutation testing are necessary to try to define the risk to offspring in families in which the phenotype may not always be clearly evident.

CASE STUDY – LEROY

Leroy is a 24-year-old man who is married to Kia. The couple are planning to start a family within the next few years, but Leroy is concerned about his family history of tuberous sclerosis.

Leroy has a sister who has been cared for all her life in a home for children with severe mental retardation. She was diagnosed with tuberous sclerosis at the age of 6 years. She has many of the typical features of the condition: epilepsy, a large shagreen patch on her back, ungual fibromas on her toenails and the typical facial 'rash' known as adenoma sebaceum. No genetic testing has been done and it would not be possible to obtain informed consent from the sister for such testing.

The couple are referred to the genetic counselor, who takes a family history. The family tree records that Leroy has a cousin on his mother's side who died young. He was in an institution and the cause of death is not known, but Leroy thinks he was 'retarded'. Leroy's parents are still living and are both in good health.

The counselor arranges a clinic appointment for the family. Leroy's parents do not want their daughter to get upset but agree to be examined. They are not aware of any signs of tuberous sclerosis in themselves. However, Leroy's mother does have one ungual fibroma, and examination under ultraviolet light (using a Wood's lamp) reveals that she has a number of depigmented patches of skin. Leroy is also examined carefully and does not have any detectable signs of tuberous sclerosis. A CT scan of his brain is ordered, and the results for this are also normal. Genetic testing is not feasible for Leroy because there are a huge range of mutations in the TSC1 and TSC2 genes and definitive testing is not practical without knowing what the mutation is that is carried by his affected sister.

Leroy is told that the risk of him having a child with tuberous sclerosis is very low as he has no phenotypic features, but the risk cannot be completely excluded because of the variability in expression of the disease. Based on this advice, Leroy and Kia decide that they will have a child.

Autosomal recessive inheritance pattern   An autosomal recessive condition is manifested when a person inherits mutations in both copies of a particular gene. All individuals carry recessive gene mutations, but these do not usually have any adverse effect on the health of the carrier, because the normal or 'wild-type' gene on the other allele is sufficient to ensure adequate gene function. In some cases, recessive gene mutations have been shown to enhance the health status of the carrier. For example, being a carrier for sickle-cell disease (SCD) confers some protection against malaria (Hebbel, 2003), and being a SCD carrier in a geographical region where malaria is endemic is therefore an advantage.

In each pregnancy, parents who are both carriers of a recessive condition have a 25% chance of having a child who is affected by the condition. Figure 3.2 shows the possible outcomes in each pregnancy. When counseling a person who has a sibling with the condition, it should be remembered that the person has a 2/3 chance of being a carrier as one possibility (i.e. being affected) has been eliminated.

Due to the carrier risk in the population being low for most recessive conditions, the risks for relatives of the affected person (other than his or her parents) having affected children is usually low. However, clients must always be asked specifically about consanguinity, as this increases the chance that both partners will be carriers of the same mutation. Figure 3.3 is a sample of a letter sent to a client after consultation, explaining their low risk.

Figure 3.2 Diagram of potential outcomes of each pregnancy in a recessive pattern

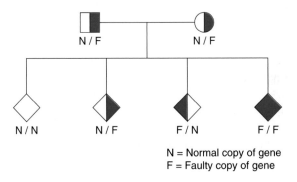

N = Normal copy of gene
F = Faulty copy of gene

Figure 3.3 **Sample letter**

Dear Roger and Mary

It was a pleasure to meet you in the genetic clinic last week, where we discussed your concerns about the family history of cystic fibrosis. I said when we met that I would write to summarize our discussion.

You were referred to the genetic clinic by Dr Brown, after you told her that you want to have a child of your own, but that you are worried because Mary's brother, Mark, has cystic fibrosis. I am sorry to hear that Mark has been very unwell lately. Apart from Mark, there is no-one else on either side of the family with cystic fibrosis. I asked whether you could have been related before you married, but you thought this was very unlikely as Roger's family is originally from Wales and Mary's parents were Italian.

I explained that cystic fibrosis is what is termed an 'autosomal recessive' condition. This means that a child who has the condition has inherited two faulty copies of a particular gene, one from each parent. The gene involved in cystic fibrosis is one that helps control the levels of salt and water in some of the body tissues. The secretions of the lungs are very thick and this makes the person who has cystic fibrosis more prone to chest infections. People who have cystic fibrosis also have problems digesting some foods and need to take additional enzymes with their food to aid digestion. About 1 in 20 people in our population are carriers of cystic fibrosis. This means that they have one faulty copy of the gene and one normal copy. Having one faulty copy of the gene usually has no effect on the health of the carrier as one working copy of the gene is enough to ensure normal function. However, if a carrier has a child with another person who is a carrier, the baby may inherit a faulty copy of the gene from both parents. The chance that this will occur in any pregnancy is a chance of 1 in 4, or 25% if both parents are carriers.

We can assume that both Mary's parents are carriers of cystic fibrosis. Mary does not have cystic fibrosis, but she has a 2 in 3 chance of being a carrier of the condition. Roger has no family history of cystic fibrosis, so his chance of being a carrier is the same as for others in the general population — a chance of 1 in 20. The chance of you having child with cystic fibrosis is $\frac{2}{3} \times \frac{1}{20} \times \frac{1}{4} = \frac{1}{120}$, or less than 1%.

We did discuss genetic carrier testing to try and refine this risk for you and you were both going to think about that before we meet again next week. In the meantime, I will ask the laboratory to send me your parents' test results to help us interpret any future test you may have. I am grateful to your parents for sending me the signed consent forms to enable me to do this.

I look forward to meeting you both again next week.

Yours sincerely

A. Practitioner

cc Family doctor

X-linked recessive inheritance pattern   When dealing with an X-linked recessive pattern, risks will differ according to the sex of the offspring. In addition, the parent may be an affected male (such as a man with X-linked retinitis pigmentosa) or a carrier female (such as a woman carrying Duchenne muscular dystrophy).

If the female parent is a carrier, her daughters have a 50% chance of also being carriers, and her sons have a 50% chance of being affected by the condition. Affected males will not have children affected by the condition. As they pass an X chromosome to their daughters, and they have only one copy of the X chromosome, all daughters will be 'obligate carriers'. Sons do not inherit an X chromosome from their father and will therefore neither be affected nor carriers.

X-linked dominant inheritance pattern   Females who inherit an X-linked dominant condition will be affected by the condition, as will males. However, the male will often have a more severe form of the disease, and some X-linked dominant conditions are lethal to the male fetus. For this reason, a disproportionate number of females may be born in these families. In each pregnancy in which the condition is lethal to males, the chance of a child inheriting the mutation is 1 in 2, but the chance of a liveborn child being affected will be 1 in 3.

Y-linked inheritance pattern   There are very few conditions that are identified as Y-linked, but those that are can only be passed on by a male parent, and all of the male children will be affected.

Risk estimates based on empirical data

There are many situations in which the inheritance pattern is not as clear as in the examples given above. In some cases, the actual genetic basis of the condition is still not known, or the condition may be multifactorial. Empirical data are then used to offer the family an idea of the level of risk. A common example of the use of empirical data to provide the family with a recurrence risk occurs when a couple have had a child with a neural tube defect. The empirical risk based on research studies (Harper, 2004) indicates that 4% of couples will have a second fetus affected by a neural tube defect after the diagnosis of this condition in a previous pregnancy. Therefore, this figure should be used in counseling a couple who have conceived a child with spina bifida (myelocele or meningo-myelocele) or anencephaly when explaining the recurrence risk to them. However, research has also shown that maternal folic acid supplements taken before and during the pregnancy reduce this recurrence risk to a rate of about 1–2% (Centers for Disease Control and Prevention, 2004). Both pieces of information are vital to enable the couple to decide whether they wish to have another pregnancy, and also whether the mother wishes to take folic acid supplements to minimize the risk. Another common example of recurrence risk based on empirical research is the risk of having a child with Down syndrome, according to maternal age.

Bayesian calculations

The Bayes theorem is attributed to Thomas Bayes, an eighteenth-century nonconformist minister. However, there is some dispute as to whether he did originally devise the calculation. Bayes' theorem can be used to calculate the recurrence risk of a genetic condition for a particular family member (possibly a fetus). At least two types of information are used within the calculation to refine and individualize the genetic recurrence risk.

The calculation is based on both the prior information (usually the risk based on the family history and known inheritance pattern) and conditional probability (such as information obtained by genetic testing or other investigations). The calculation results in a likelihood ratio. This ratio can be converted into an expression of risk as a percentage or fraction. The Bayesian calculation can be used to assess the risk of being affected or the risk of being a carrier. For brevity, the calculation will be explained using as a method of assessing carrier risk, but the calculation can also be used in the same way to assess the risk of a person being affected by a condition.

The steps for calculating the odds ratio are:

Step 1 – Prior probability   Identify the prior probability of the person being a carrier of the condition. This is usually based on the known inheritance pattern and the individual's position in the family.

The risk of the individual NOT being a carrier is also required.

For example, the chance that the daughter of a carrier of X-linked retinitis pigmentosa (XLRP) is a carrier herself is 1 in 2 and the risk of her not being a carrier is also 1 in 2.

Column A	Column B
Chance of being a carrier	Chance of not being a carrier
½	½

Step 2 – Conditional probability   Identify the probability of the test result or other information occurring if the person is a carrier, and the probability of it occurring if the person is not a carrier.

For example, the daughter of the XLRP carrier has two healthy sons who have no signs of the condition. The chance of this happening if she is a carrier is $\frac{1}{2} \times \frac{1}{2} = \frac{1}{4}$. The chance that she will have 2 unaffected sons if she is not a carrier is 1.

Column A
Chance of 2 unaffected sons if
a carrier
$\frac{1}{2} \times \frac{1}{2} = \frac{1}{4}$

Column B
Chance of 2 unaffected sons if not
a carrier
1

Step 3 – Joint probability   The joint probability combines the prior and conditional probability information.

To calculate the joint probability, the figures in Column A are multiplied together and the figures in Column B are multiplied together.

Column A
$\frac{1}{2} \times \frac{1}{4} = \frac{1}{8}$

Column B
$\frac{1}{2} \times 1 = \frac{1}{2}$

To make the odds ratio easier to calculate, the fractions can be written using the same denominator in both columns.

Column A
$\frac{1}{8}$

Column B
$\frac{4}{8}$

The odds of the woman's daughter being a carrier are $\frac{1}{8}$ to $\frac{4}{8}$, or more simply, 1 to 4 (1:4).

Step 4 – Phrasing the odds ratio as a risk   The odds ratio can be converted to a percentage risk if this is the preferred way of presenting the carrier risk.

Note that odds of 1 to 4 means 1 chance to 4 chances, and this equates to a chance of 1 in 5.

In the example, 1:4 odds = $\frac{1}{5}$ risk = 20%

CASE STUDY – MARGAUX

Margaux is a 56-year-old woman who has a strong family history of cancer. Three maternal aunts and her maternal grandmother were affected by breast cancer. However, Margaux's mother is now 92 years old and has not had any form of cancer.

Affected family members have a BRCA2 mutation that has a 70% penetrance rate for breast cancer.

Margaux does not want to ask her elderly mother to be tested for a BRCA2 gene mutation, nor does she necessarily want to be tested herself. She asks about her risk of developing breast cancer, based on the family history. A Bayesian calculation is used to assess her mother's chance of having inherited the familial mutation.

	If a mutation carrier	If not a mutation carrier
Prior probability (what is the chance that Margaux's mother inherited the mutation, based on a dominant inheritance pattern?)	$\frac{1}{2}$	$\frac{1}{2}$
Conditional probability (what is the chance that Margaux's mother inherited the mutation, but has had no history of breast cancer?)	$\frac{3}{10}$	1
Joint probability	$\frac{1}{2} \times \frac{3}{10} = \frac{3}{20}$	$\frac{1}{2} \times 1 = \frac{1}{2}$
Express using same denominator in each column	$\frac{3}{20}$	$\frac{10}{20}$
Odds ratio (odds that Margaux's mother inherited the mutation)		3:10
Risk expressed as a fraction or percentage (chance that Margaux's mother inherited the mutation)		$\frac{3}{13} = 23\%$

As the chance that Margaux's mother has inherited the BRCA2 mutation that was identified in other family members is $\frac{3}{13}$, Margaux's own risk is $\frac{3}{26}$ or 11%. On this basis, Margaux decides not to have any further testing.

CASE STUDY – MARVIN

Marvin's father, Lee, has neurofibromatosis (NF). Lee is concerned about his son, and asks the geneticist to examine him for signs of NF. Marvin has no café au lait patches or any other sign of NF.

Marvin is now 5 years old. According to a recent publication, the geneticist is aware that 95% of children with NF will have café au lait patches by the age of 5 years.

	If affected	If not affected
Prior probability (what is the chance that Marvin inherited NF, based on a dominant inheritance pattern?)	$\frac{1}{2}$	$\frac{1}{2}$
Conditional probability (what is the chance that Marvin inherited NF but has no café au lait patches by the age of 5 years?)	$\frac{1}{20}$	1
Joint probability	$\frac{1}{2} \times \frac{1}{20}$ $= \frac{1}{40}$	$\frac{1}{2} \times 1 = \frac{1}{2}$
Express using same denominator in each column	$\frac{1}{40}$	$\frac{20}{40}$
Odds ratio (odds that Marvin inherited NF)		1:20
Risk expressed as a fraction or percentage (chance that Marvin has inherited NF)		$\frac{1}{21}$ $= < 5\%$

The Hardy–Weinberg equation

The Hardy–Weinberg equation derives from population genetics and is used to calculate the chance of a member of the general population carrying a particular autosomal recessive condition. This is important when a couple are seeking genetic advice because one partner has a family history of a recessive condition. Although the risks to offspring in this situation are generally low, it is necessary to know the carrier risk for both partners in order to calculate the chance of a child of the couple inheriting the condition.

To use the equation to determine the number of people in a given population who are heterozygotes (carriers of a mutated gene), the incidence of the condition (number of people in the population affected) must be known. As this can vary enormously across populations, care must be used to ensure that the figures are relevant to the geographical region and ethnic origin(s) of the couple.

The equation states that:

$$p^2 + 2pq + q^2 = 1,$$ where p is the normal allele and q is the mutated allele of a particular gene.

The frequency of homozygotes (q^2) (number of affected people) in a population is used to calculate the carrier frequency for the condition (2pq).

For practical purposes, because the population sample is so large the frequency of the normal allele (p) is regarded as 1. Therefore, the carrier rate in the population (frequency of heterozygotes) is 2q.

Using the equation

Case example – James and Karen:

James is the partner of Karen. Karen is a carrier of spinal muscular atrophy (SMA) (Werdnig–Hoffman type).

In the region in which James and Karen live, 1 in 10,000 children are born with this type of SMA.

Calculation

$p^2 + 2pq + q^2 = 1$

$q^2 = \frac{1}{10,000}$

The square root of $\frac{1}{10,000} = \frac{1}{100}$

$2pq = 2 \times 1 \times \frac{1}{100} = \frac{1}{50}$

James' carrier risk is 1 in 50.

The risk of James and Karen having a child with SMA is 1 (Karen's carrier risk) $\times \frac{1}{50}$ (James' carrier risk) $\times \frac{1}{4}$ (risk of autosomal recessive condition if both parents are carriers) $= \frac{1}{200}$.

The equation is used in later chapters (Chapters 9 and 10) to demonstrate calculation of carrier risk in specific conditions.

3.4 Prenatal testing

Prenatal testing requires considerable input from the healthcare practitioner with respect to:

- the psychological preparation and support of the couple;
- liaison between the couple, the genetics department, the obstetric team
- and the laboratory;
- interpretation and explanation of the genetic results.

Psychological preparation and support of the couple

Preparation for prenatal testing preferably starts well before conception. The practical possibilities for testing should be investigated by the practitioner so that a full range of options can be presented to the couple. For example, it may be possible to obtain a sample using chorionic-villus biopsy or amniocentesis. In addition, it is usually necessary for molecular genetic testing to obtain samples from affected family members to confirm the mutation or undertake linkage analysis to define the most useful combination of polymorphic markers. In some cases, prenatal testing may not be feasible and the couple should be made aware of this before starting a pregnancy. The decision to have prenatal diagnostic testing is usually made because a couple consider the condition to be very serious. This type of testing is not commonly used for adult-onset conditions.

Counseling for prenatal testing should always include discussion of the management of the pregnancy in the event of adverse results. Termination of pregnancy should not be a condition of prenatal testing, but if the couple would not consider termination of pregnancy, they need to be aware of the risks to the pregnancy of having an invasive test. The risk of miscarriage due to the procedure varies according to the operator, but is usually between 0.5% and 1% (Nanal *et al.*, 2003). Each obstetric department should have figures relating to the pregnancy loss rates for that center that can be used for accurate counseling.

The period of time leading up to the test and awaiting the results is often stressful for the couple and additional support may be appropriate in the form of counseling sessions or telephone calls. At this time, they may have limited support from other sources as many couples choose to keep the pregnancy secret until they are certain of the test outcome.

Following the results, couples may wish to have further discussions about the future of the pregnancy if the fetus has been shown to be affected. Continuing support after termination or contact after the birth of the baby is usually appreciated and helps to maintain the links between the practitioner and the client, which is a valuable part of any healthcare service. If the pregnancy is terminated, it is particularly meaningful to contact the family around the time the fetus would have been born as this is often a very distressing time for the family.

In some cases, parents may receive a result indicating that the fetus is not affected, but is a carrier of the condition. This result needs careful explanation to ensure that the parents are not unduly anxious about the child. Plans for long-term follow-up to offer genetic counseling to the child when he or she is at an appropriate age should be made.

Liaison between the couple and the healthcare professionals

The genetic healthcare practitioner is often the first professional to be contacted by the mother after a pregnancy is confirmed, particularly if extensive discussion has already taken place. Although the organization of services differs from center to center, the practitioner will often be required to ensure that:

- the couple are offered a further opportunity to discuss prenatal testing if they wish;
- an ultrasound dating scan has been arranged so that the timing of tests can be planned;
- the laboratory staff are aware of the pregnancy and the date on which samples will be taken;
- the obstetric/gynecology team are aware of the pregnancy and the type of testing planned;
- there is provision for termination of pregnancy without undue delay should the results be adverse and the couple decide not to continue the pregnancy;
- there is a mechanism to ensure fetal samples are sent for analysis after termination of pregnancy, if appropriate;
- if the pregnancy continues, there is a mechanism to report the outcome to the genetics service.

Interpretation and explanation of results

In most cases, this will be straightforward, and the potential results will have been discussed with the couple before they consent to testing. However, there are occasions when the test results are not clear, unexpected findings arise or the prognosis for the fetus cannot be well defined. Clients may wish to take the opportunity to discuss the condition and potential prognosis with other clinicians such as a pediatrician before coming to a decision.

3.5 Predictive testing

The term predictive testing generally refers to genetic testing for adult-onset conditions prior to the onset of signs or symptoms. The client has the right to request such a test and make an autonomous, informed decision about the appropriateness of the test for themselves. It is the responsibility of the practitioner to ensure that the client is properly informed, has considered the implications and is as prepared as possible for dealing with the result. Protocols for such testing were first devised for use in predictive testing for neurological conditions such as Huntington disease. These protocols include pre-test counseling to:

- discuss the process of testing;
- enable the client to explore the implications of both positive or negative results on his or her own life;
- explore the implications of test results for key aspects of life, such as relationships, reproductive plans, employment, financial issues, and insurance;
- explore with the client the effects of the result on others, including other family members at risk;
- discuss ways in which the client has dealt with other stressful situations and facilitate the client in devising methods of dealing with the news;
- ensure that the client has appropriate support while preparing for the test and dealing with the result.

In some centers, a psychologist or psychiatrist may see the client to make an assessment of the client's mental state. This should always be done before testing if the practitioner has any concerns about the client's past medical history or current mental state. Clients seeking predictive testing are usually seen by experienced practitioners who have demonstrated competence in this area of practice.

The results should be given in person at a pre-agreed time and place. After the results are given, regular contact to offer support while the client makes the adjustment to their status should be offered. It is just as important to offer support to those who have a negative result as to those who have a positive mutation test. Many people at risk convince themselves that they will develop the condition as a psychological defense, so it can require a significant adjustment to accept that they have a future that will be free of the disease. These clients may also suffer from 'survivor guilt', especially if siblings are affected or still at risk.

For those individuals who have been found to have inherited the mutation, a system of clinical surveillance can be instituted if appropriate, and long-term support offered.

There is a further discussion of predictive testing in Chapter 8.

3.6 Testing children

Ethical issues about the genetic testing of children are discussed in the next chapter. However, it is appropriate to test children who are symptomatic, to confirm a diagnosis, or when medical interventions or surveillance before

---

CASE STUDY – BEN

Ben is 8 years old and has moderate learning difficulties. He has some mild dysmorphic features, including a large depigmented patch on one side of his abdomen and has frequent epileptic seizures. Ben had a blood sample taken for karyotyping when he was anesthetized for insertion of Grommets (for drainage of fluid in the ear) when he was 6 years old. The karyotype was reported as normal.

Ben's geneticist believes that Ben may have chromosomal mosaicism but this was not detected in the lymphocyte chromosome study. She decides to take a skin sample for fibroblast chromosome studies.

The nurse explains to Ben that the doctor wants to take a little piece of skin to test, to try and find out why he has fits. She explains that she will put some cream on his arm and wrap some plastic around it while he plays in the playroom for a while. When he comes back to the room his lower arm will be washed with 'cold, pink water' and the doctor will take a little piece of skin and put it into a jar.

Ben looks worried when he comes back into the room, but the nurse continues to tell him what is happening and he is fascinated by the sterile strips that are used to close the tiny wound. Ben's teddy bear is also given a sterile strip across his arm.

Four weeks later, the result is received from the cytogenetics laboratory. It indicates that Ben has mosaicism. Some of his cells have two normal copies of chromosome 8, but a small proportion have only one copy of that chromosome.

adulthood could be beneficial to the child. The purpose of the test and the way the sample will be obtained is explained to the child in terminology that is appropriate for the child's age and level of understanding. Where blood or skin samples are required, it is helpful to use local anesthetic preparations to minimize pain and distress.

Parents are naturally anxious to reduce anxiety in the child and may ask the practitioner to collude in giving a false reason for the testing. This has to be dealt with individually, but the practitioner should not be involved in any activity that would compromise their professional integrity or create distrust between the practitioner and the child.

3.7 Organization of clinical screening

The role of the genetic healthcare practitioner may involve organization of clinical screening or surveillance for clients. Some of the tests that are used for clinical surveillance in a sample of genetic conditions are listed in Table 3.1.

Organization of clinical tests for surveillance of clients is preceded by discussion with the colleagues who will be providing the surveillance, to agree on protocols for screening according to the level of risk. The purpose and type of screening is discussed fully with the client before the referral is made. It may be necessary to monitor the frequency and results of screening to ensure that the organizational system is working effectively.

Table 3.1 Tests used for clinical surveillance

Investigation	Relevant conditions
Colonoscopy	Familial adenomatous polyposis (FAP) Hereditary non-polyposis colon cancer (HNPCC) Peutz Jegher syndrome
Mammogram	Familial breast and ovarian cancer
Echocardiogram	Marfan syndrome Hypertropic cardiomyopathy
Renal ultrasound	Adult polycystic kidney disease von Hippel Lindau syndrome (VHL) Tuberous sclerosis
Serum iron studies	Hemochromatosis

3.8 Managing rare complex conditions

It has already been mentioned that genetics services are offered both in generic centers and in disease-based facilities. In generic centers, management of disease may not be a key purpose of the service, and clients will be referred to other appropriate clinical teams for treatment and management. However, in some complex disorders there may be numerous clinical teams involved in the care of a client. In that case, the genetic team may take on the role of coordinating care. For example, the genetic service may run a clinic for families with von Hippel Lindau disease to organize screening for abnormalities of the brain, eyes, kidneys and liver. The treatment of the condition is still, however, usually carried out by the relevant specialist.

In some highly specialized centers, the genetic healthcare practitioner may be involved in research trials of gene therapy and/or pharmacogenomics.

3.9 Maintaining clinical registers/registries

The term register (or registry) is used in genetic healthcare in a number of different contexts. In some cases, it means a simple database of names, to record the families affected by a specific condition for ease of retrieval (e.g. to contact families when new treatment or testing options are available). Clinical registers can also be dynamic tools that are used to facilitate contact with families on a regular basis. In this case, the use of the register ensures that appropriate services are being offered to family members. For example, it may be used to contact potential carriers of a condition to offer carrier testing in early adulthood. In either case, the clinical register is a form of medical record and the client's confidentiality must therefore be protected and unauthorized access to the register prevented.

3.10 Conclusion

In this chapter, the practical tools that are required for basic genetic healthcare have been described. However, development of practical skills must be considered in the context of the ethical and legal boundaries to practice, and these are considered in the next chapter.

Study questions

Read the case study of Leroy and Kia again (p. 41)

Task A
(a) Using sources such as the website for the Tuberous Sclerosis Association, make a list of the potential signs of tuberous sclerosis that could provide a clue as to whether an individual has inherited the mutation.
(b) Leroy had a brain scan. What type of finding would have indicated he had tuberous sclerosis and why is this finding seen in patients with this condition?

Task B
Samples for DNA analysis were not taken from Leroy's sister Zaina. She was not able to give informed consent and her parents did not want samples taken from her.

(a) Suggest reasons for the parents having this view?
(b) In your country, is it legally permissible to take a sample for DNA testing from an adult who cannot give informed consent?
(c) Discuss in a group how you might ensure that you work ethically in this family, upholding the rights of both Leroy and Zaina.

References

Bennett, R.L. (1999) *The Practical Guide to the Genetic Family History*. Wiley-Liss, New York.

Centers for Disease Control and Prevention (CDC) (2004) Spina bifida and anencephaly before and after folic acid mandate – United States, 1995–1996 and 1999–2000. *MMWR Morb. Mortal Wkly Rep.* **53**: 362–365.

Harper, P.S. (2004) *Practical Genetic Counselling*, 6th Edn. Oxford University Press, Oxford.

Hebbel, R.P. (2003) Sickle hemoglobin instability: a mechanism for malarial protection. *Redox Rep.* **8**: 238–240.

Hockley, J. (2000) Psychosocial aspects in palliative care – communicating with the patient and family. *Acta Oncol.* **39**: 905–910.

Keiley, M.K., Dolbin, M., Hill, J., Karuppaswamy, N., Liu, T., Natrajan, R., Poulsen, S., Robbins, N. and Robinson, P. (2002) The cultural genogram: experiences from within a marriage and family therapy training program. *J. Marital Fam. Ther.* **28**: 165–178.

Nanal, R., Kyle, P. and Soothill, P.W. (2003) A classification of pregnancy losses after invasive prenatal diagnostic procedures: an approach to allow comparison of units with a different case mix. *Prenat. Diagn.* **23**: 488–492.

Skirton, H. (2001) The client's perspective of genetic counseling – a grounded theory approach. *J. Gen. Counsel.* **10**: 311–330.

United States Department of Health and Human Services. Office for Civil Rights – HIPPA. Medical Privacy – National standards to protect the privacy of personal health information. [Accessed July 26, 2004] http://www.hhs.gov/ocr/hipaa/finalreg.html

van Dijk, S., Otten, W., Zoeteweij, M.W., Timmermans, D.R., van Asperen, C.J., Breuning, M.H., Tollenaar, R.A. and Kievit, J. (2003) Genetic counselling and the intention to undergo prophylactic mastectomy: effects of a breast cancer risk assessment. *Br. J. Cancer* **88**: 1675–1681.

Further resources

Barr, O. and Millar, R. (2003) Parents of children with intellectual disabilities: their expectations and experience of genetic counselling. *J. Applied Res. Intellect. Disabil.* **16**: 189–204.

Contact a Family (2004) Working with families affected by a disability or health condition from pregnancy to pre-school – a support pack for health professionals. [Accessed March 7, 2005] http://www.cafamily.org.uk/HealthSupportPack/pdf

Council for Responsible Genetics [Accessed March 7, 2005]. http://www.genelaw.info/

Emery, J. (2001) Is informed choice in genetic testing a different breed of informed decision-making? A discussion paper. *Health Expect.* **4**: 81–86.

NSGC. http://www.nsgc.org/consumer/familytree/index.asp

South West Thames Regional Genetics Service. http://www.genetics-swt.org/fhist.htm

Tassicker, R., Savulescu, J., Skene, L., Marshall, P., Fizgerald, L. and Delatycki, M.B. (2003) Prenatal diagnosis requests for Huntington's disease when the father is at risk and does not want to know his genetic status: clinical, legal and ethical viewpoints. *BMJ* **326**: 331–333.

4 Working professionally in genetic healthcare

4.1 Professionalism

Practitioners in genetic healthcare – whatever their basic training and background – regard themselves as professionals. Two of the basic requirements for a profession are self-regulation and the adherence by practitioners to a code of ethics. Although working in an ethical way will often be a matter of routine and common sense, there are situations in which the correct course of action is unclear, and a code of practice that is agreed by the profession is needed as guidance. In this chapter, different ethical codes and potentially difficult situations will be discussed.

4.2 Working ethically

Healthcare practice is governed by codes of ethics. The use of ethical principles in healthcare practice is necessary to protect those individuals who use the service. Discrimination against those with a genetic condition or risk of an inherited disease has existed in many countries (Kevles, 1999); this has been aided by the collusion of health professionals. One example of the national implementation of eugenic policies was in Sweden, where the sterilization of individuals who were considered unfit to breed occurred well into the second half of the twentieth century. Eugenic policies have been instituted in many countries, but the Eugenics movement originated in the UK under the auspices of Francis Galton (Blacker, 1945) and was primarily concerned with improving the fitness of the country's citizens. Positive eugenics aims to encourage those individuals who are most able and healthy in society to breed, whereas negative eugenics seeks to prohibit or dissuade the 'unfit' from producing children. Eugenic policies obviously raise significant ethical questions –

such as the ways in which 'fitness' is judged and the potential infringements of the basic human right to have a family, which is enshrined in the European Convention on Human Rights (http://conventions.coe.int/treaty/).

There is no one perfect 'code of ethics' but a number of codes that are based on different philosophical foundations. For example, virtue ethics is based on a moral code that relies on the good character of the practitioner. Various moral virtues, such as kindness, concern for others and honesty, form the basis of the code. Utilitarian ethics is based on the premise that any action should result in the greatest good for the largest number of people. An action can be defended if it results in an overall benefit for the majority. There may be practical difficulties with both these approaches for the genetic healthcare practitioner. It could be argued that acting virtuously towards a client may, in fact, reduce their autonomy, if acting kindly means shielding the client from unpalatable news. On the other hand, from some perspectives doing the greatest good might mean depriving a disabled person of life to make more health resources available to others in the community. Obviously, neither of these codes address the needs of the practitioner in this setting, and other sources of support for ethical decision-making and practice must be used.

The Hippocratic Oath as a basis for ethical practice in genetic healthcare

Although the Hippocratic Oath was devised to guide the medical profession, it is generally regarded by health professionals from other disciplines as a viable guide for their own practice. The Oath states that those who practice must have special knowledge and ability; in modern terms, this means the practitioner is obliged to practice work within the boundaries of his or her own knowledge and experience. The other basic concept included in the Hippocratic Oath is that the practitioner must always act in the best interests of the client. However, although this may be relatively straightforward when working with an individual, the 'client' in genetic settings is often the family. The practitioner may be caring for more then one member of a family and there may be genuine conflicts of interest between family members. This may even relate to conflict between an adult individual and a fetus.

Principles of ethical practice

There are four main principles that can be used both as general guides in practice and to support decision-making when there are conflicts of interest (Halsey Lea et al., 1998).

Non-maleficence The principle of non-maleficence decrees that the practitioner should do no harm to the client. In genetic healthcare, it could be said that the potential for harm is less than in other settings. For example, the practitioner is not usually undertaking invasive procedures. However, there are opportunities for harm, some of which may occur when:

- the client's confidentiality is breached;
- the client feels coerced into making an inappropriate decision;
- the practitioner is not adequately trained for the work and provides the client with incorrect or inadequate information;
- the practitioner is unwell or not performing appropriately.

Beneficence Beneficence broadly means 'to do good' and the principle states that the practitioner engages in activity that is advantageous to the client. Although this may seem straightforward, it is important that the benefit is judged from the client's perspective; therefore, the client's needs and preferences must be respected.

Autonomy Autonomy is a legal as well as an ethical principle. In law, a person's autonomy is protected, as illustrated by the need for informed consent for genetic testing in most countries. In ethical terms, enabling self-determination is underpinned by respecting the individual's right and ability to make decisions. Using a non-directive approach in genetic healthcare is one way of respecting autonomy.

Justice All clients are entitled to justice in respect of their genetic healthcare. The use of the term justice in an ethics context generally relates to justice in a broad form, such as ensuring equity of care and employing anti-discriminatory polices (Singleton and McLaren, 1995). However, justice on an individual basis may mean ensuring that the practitioner does not discriminate against a client because of personal prejudices. In practical terms, it is clear that all clients cannot always be offered the same level of service, but differences in service should be based on fair use of resources and carefully devised policies. For example, it may be decided by members of a genetics team that has a restricted budget for genetic testing that resources will not be used for testing in those families in which the test results could make no difference to health outcomes. In one scenario, testing would be offered such that the results would help identify those individuals who should be offered clinical screening, whereas tests might not be funded for those whose status is already phenotypically known.

Making ethical decisions In professional practice, situations will arise in which, despite using the ethical principles above, the safest approach is not obvious (Tassicker *et al.*, 2003). As a general heuristic, avoiding harm should take precedence over the other principles. However, an individual should not undertake these decisions alone. Discussion with experienced clinical colleagues can help to clarify the situation, supported by discussion with experts in health ethics (Schmerler, 1998; Schnieder *et al.*, 2000). Discussion groups, such as the 'GENETHICS' club in the UK, have been specifically formed to enable structured discussion of ethically challenging cases between professionals, thereby helping to guide good practice. Ethics committees in some US healthcare institutions serve a similar purpose.

Chadwick (1999) writes that new genetic technology has altered the basis of ethical practice, changing the notion of choice and responsibility. Autonomous choice places on the client a duty to know more about their own identity and beliefs. This, in turn, places responsibility on the practitioner to facilitate discussion that includes these concepts.

Consent   A particular aspect of ethical practice concerns gaining consent for procedures. The key aspects of consent are that the person understands the nature and risks of the procedure to which they are consenting, and that the person gives consent without coercion.

In genetic healthcare settings, consent most often relates to the following:

1. **Taking a family history.** Consent can generally be assumed if the proband provides the information requested, providing that the process and reason for taking the pedigree have been explained. However, when using the pedigree to counsel other family members, the confidentiality of the original proband must be respected. For this reason, it may be appropriate to take a new pedigree when seeing a different branch of the family.
2. **Obtaining specific medical history from the proband and/or other relatives.** It is frequently necessary to request medical notes on the proband in order to advise him or her properly. However, consent must be sought to view or request medical records. The purpose of viewing records of other family members must be explained to them and written consent obtained.
3. **Obtaining blood or tissue samples.** Permission to take a sample must be explicitly given by the client. This is sometimes written consent, but if the procedure has been explained the co-operation of the client in giving the sample is usually deemed to be evidence of consent. For example, if a client lifts his sleeve and presents his arm after being asked to consent to a blood sample, this would be evidence that the client has given consent.

4. **Performing genetic tests.** The exact nature of the tests and the implications of the results must be explained to the client (Emery, 2001). It is good practice to give the client written information as well as a verbal explanation, and written evidence of consent must be recorded. Risks associated with genetic testing might include the discovery of false paternity; this should be mentioned if there is a known possibility. Other aspects of consent for genetic tests include whether consent is given for the sample to be stored and the possible outcomes of the test. Separate consent should be obtained for use of the sample in research and to share the results with relatives in the process of their own testing.

Confidentiality Confidentiality of personal information is a basic tenet of healthcare and is considered so important to the rights of the client that it is enshrined by law in many countries. However, there may be provision under some statutes for the healthcare professional to disclose the client's confidential medical information, if not disclosing would result in serious but avoidable harm to others. This is the case in UK law. For example, confidentiality might be breached if a person had a serious infectious disease that was putting others in the community at risk.

In a genetic healthcare setting, the situation may be complex, as the information about the genetic structure of one individual may (and often does) have implications for other family members. Where this occurs, the proband is usually encouraged to share the information with relatives who may be affected, especially if screening or treatment is available that would reduce the health risk. It is usual to offer support in the form of written information that can be given to relatives and also contact details so that they can seek more information and guidance from the genetics team if they wish.

When an individual refuses to share information with relatives, there is always an underlying reason that might not be obvious to the practitioner. The situation is rarely urgent, and effort spent in gaining the proband's confidence and allowing time for psychological adjustment to their status can often be helpful in enabling the proband to share the information. However, this is not always the case and then the decision about whether to break confidentiality may arise. This should always be discussed with experienced colleagues, but the decision tree outlined in Figure 4.1 may be helpful.

Figure 4.1 Ethical decision tree

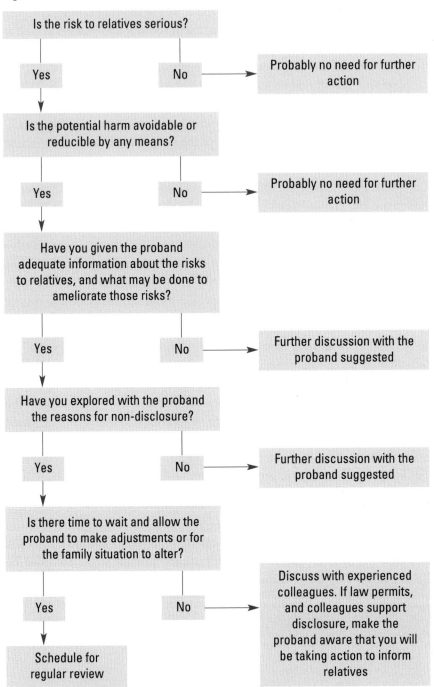

Document the decision-making process appropriately, giving reasons for the decisions and recording discussions with colleagues.

4.3 Working legally

Within all countries, there are laws and statutes that govern the provision of healthcare. In addition to conforming to the legal requirements for practice, statutory registration as a practitioner may be mandatory. For example, in the USA genetics nurses must be licensed within the state where they practice, and nurses in the UK and other European countries are registered with a statutory body that governs professional standards.

It is not possible in this text to describe the legal boundaries of practice, as these differ between countries and even between states. This section is included only to bring the issue of legal responsibility to the practitioner and to advise that information is sought through professional bodies representing the profession in the country of practice.

4.4 Boundaries of practice

The boundaries of professional practice may be defined by law for each professional group. However, in countries in which the provision of genetic healthcare is relatively new, the role of the practitioner may be undefined. Whether or not the duties and responsibilities of the genetic health professional are well regulated, it is an integral part of ethical practice to work only within the realms of one's expertise. The practitioner should be trained appropriately for the roles he or she performs, and ultimately the onus is on the professional to draw the boundaries of practice. This can be challenging for a learner in the field, and an experienced mentor or supervisor should be approached for ongoing guidance.

4.5 Using supervision

The term supervision is used in two slightly different contexts in genetic healthcare settings. The use of a counseling supervisor is discussed in Chapter 5. However, in addition to counseling supervision, the practitioner should also use a clinical supervisor. The clinical supervisor is preferably an experienced practitioner from the same professional background as the supervisee, but in small teams this is not always possible. The most important aspect of the supervisory relationship is trust between the persons concerned.

Regular sessions for supervision should be set aside, but as challenging situations may arise unexpectedly, the potential for *ad hoc* consultation should also exist. The purpose of clinical supervision is to:

- enable the practitioner to access the experience and knowledge of another professional on a formalized basis;
- facilitate the person being supervised to reflect on his or her own practice;
- protect the client through this method of clinical governance.

4.6 Ensuring appropriate documentation

As a health professional, the practitioner in genetics is subject to the legal requirements for documentation of contact with clients. All documents pertaining to the case should be stored in the record, including handwritten records of contact, test results, and all correspondence relating to the case.

The medical record may be produced as evidence in court and each entry should be signed and dated. Each page of the record should be headed with the client's name and identifying number.

Each personal contact with the client should be recorded:

- contemporaneously – the record is written at the time of the contact. Gaps on the page should not be left between entries, as each entry should follow contemporaneously. If a gap is left inadvertently, a line is drawn across the gap so new information cannot be inserted later;
- factually, with any direct quotes from the patient put in inverted commas;
- non-judgmentally, with statements of opinion identified as such.

A typical record of personal contact with the client, either by telephone or face-to-face, will include:

- the names of those present;
- the reason for the contact, e.g. 'Mrs Dean phoned to ask if her results were available yet';
- issues discussed;
- plan of action.

The client is usually sent a letter summarizing the discussion after any significant contact; this also serves as a record of the encounter.

Sample notes from the medical record are provided in Figure 4.2.

Figure 4.2 **Sample documentation of a clinical contact**

Name of client or family: Leon Vance DOB 2/6/1974
Genetics Record Number: CG12345

Date 1/6/2005

Mr Vance was seen in the genetic department today because of his concern about the family history of Marfan syndrome. Referred to genetic service by his family physician after his brother was diagnosed with a dilated aortic root. Brother has given signed permission for his medical notes to be viewed.

Family tree taken.

Medical history of brother: Henry Vance born 1/5/1972

Height: 6 feet 3 inches

No problems with eyes.

Dislocated shoulder on one occasion when playing football.

Felt unwell recently – referred to cardiologist – dilated aortic root diagnosed on cardiac echogram.

Leon's father and mother both tall but no significant health problems in them. He thinks they would be willing to be examined.

Leon has no children but married 6 months ago and he and his wife would like a family. They are shocked by the diagnosis in his brother – 'it just came out of the blue' and want to know:

1) Does Leon have Marfan syndrome?
2) What are risks of future children being affected?

Discussed in clinic:

Signs and symptoms of Marfan syndrome

Genetic basis of Marfan syndrome and dominant inheritance pattern

Tests available – genetic and clinical.

Leon's fears for himself and his offspring.

Agreed plan:

Arrange physical examination and echocardiogram for Leon.

Arrange for Leon's parents to be offered a physical examination and echocardiogram.

Await DNA test results on Henry. When results ready, offer testing to Leon and his parents.

See Leon with his wife to discuss prenatal diagnosis if Leon has mutation.

Signed

A Practitioner

4.7 Research evidence

Competent practitioners are able to use research evidence to ensure that clients are given current, accurate information. In a rapidly changing field, this can be challenging on a day-to-day basis, and the use of sites on the worldwide web can support foundation knowledge. The use of research and involving clients in research is covered in depth in Chapter 13.

4.8 Conclusion

The responsibilities of the professional in genetic healthcare are governed by both ethical guidelines and the law in each country. The practitioner has a duty of care to the client, and this involves being knowledgeable about the governance of his or her profession and adherence to the guidelines for the safety of clients.

Study questions

Read the following case scenario:

Janice is a genetic practitioner who has completed her basic training. She is asked to see a young man, Dean, who is concerned about his recent diagnosis of Klinefelter syndrome (46, XXY). When she meets Dean, she explains the genetic basis of the chromosome anomaly. She asks him how the diagnosis has affected him, but is disconcerted by the fact that Dean becomes very emotional about his relationship with his girlfriend, who wants to have children.

Janice does not feel that talking about the future of the relationship is within her scope of practice as a genetics counselor. She summarizes the information on Klinefelter syndrome and terminates the meeting, offering to see him again if he has further questions.

(a) Thinking about ethical principles, has Janice acted in the best interests of the client?
(b) Describe the ways in which she may have influenced Dean's well-being by her actions in this session?
(c) Describe how Janice might have acted differently, in accordance with the code of ethics of your own professional group.

References

(All references in this chapter are recommended as further reading)

Blacker, C. (circa 1945) *Eugenics in Retrospect and Prospect.* The University Press, Glasgow.

Chadwick, R. (1999) Genetics, choice and responsibility. *Health, Risks and Society* **1**: 293–300.

Emery, J. (2001) Is informed choice in genetic testing a different breed of informed decision-making? A discussion paper. *Health Expect.* **4**: 81–86.

Kevles, D.J. (1999) Eugenics and human rights. *BMJ* **319**: 435–438.

Halsey Lea, D., Jenkins, J.F. and Francomano, C.A. (1998) *Genetics in Clinical Practice. New Directions for Nursing and Health Care.* Jones and Bartlett Publishers, Sudbury.

Schmerler, S. (1998) Ethical and Legal Issues. In: *A Guide to Genetic Counseling* (eds D. Baker, J.L. Schuette and W.L. Uhlmann). Wiley-Liss, New York.

Schneider, K., Kieffer, S. and Patenaude, A. (2000) BRCA1 testing and informed consent in a woman with mild mental retardation. *J. Genet. Counsel.* **9**: 411–6.

Singleton, J. and McLaren, S. (1995) *Ethical Foundations of Health Care.* Mosby, London.

Tassicker, R., Savulescu, J., Skene, L., Marshall, P., Fitzgerald, L. and Delattycki, M.B. (2003) Prenatal diagnosis requests for Huntington's disease when the father is at risk and does not want to know his genetic status: clinical, legal and ethical viewpoints. *BMJ* **326**: 331–333.

Further resources

Campbell, A., Charlesworth, M., Gillett, G. and Jones, G. (1997) Medical Ethics. Oxford University Press, Oxford.

Canadian Association of Genetic Counselors (1999) Scope of Practice. [Accessed August 2, 2004] http://www.Cagc-accg.ca

Genethics Club. [Accessed March 3, 2005] http://www.ethox.org.uk/genethics/index.htm

Harper, P.S. and Clarke, A.J. (1997) *Genetics, Society and Clinical Practice.* BIOS Scientific Publishers, Oxford.

ISONG (1998) Statement on the scope and standards of genetics clinical nursing practice. American Nurses Publishing, Washington DC, USA.

Singleton, J. and McLaren, S. (1995) *Ethical Foundations of Health Care.* Mosby, London.

United States Department of Health and Human Services. Office for Civil Rights – HIPPA. Medical Privacy – National standards to protect the privacy of personal health information. [Accessed July 26, 2004] http://www.hhs.gov/ocr/hipaa/finalreg.html

5 Working to support families

5.1 Using counseling skills in the context of genetic counseling

Genetic counseling has been defined by the American Society for Human Genetics (ASHG) (ASHG, 1975) as a communication process. However, there are distinct differences between the practice of genetic counseling and the process of counseling. Ordinarily, counseling in a general setting involves active listening by the counselor and the use of counseling interventions to enable the client to express his or her feelings. The counselor facilitates the client to discover the focus of his or her difficulty and explore ways in which it might be managed. However, in a genetic counseling setting, information is traded between client(s) and practitioner. A family tree may be taken in a purposive way and, at appropriate times in the session, information about the genetic condition and inheritance risks can be provided for the family.

There is a demonstrable difference between the use of counseling skills and active counseling. In all healthcare settings, the use of counseling skills is helpful in enabling the client to discuss their health concerns and engage in shared decision-making about investigation and/or treatment. The core competencies for genetic healthcare professionals include eliciting the client's concerns and exploring the psychosocial influences that have relevance to the genetics counseling for each family (AGNC, 2004). However, there will be sessions when the need to use counseling skills is more apparent, such as when the client is making difficult decisions or during periods of adjustment to changed circumstances. Active counseling may be undertaken by genetics counselors who are appropriately trained to assist the client when the psychosocial issues are impeding adjustment to their genetic situation and therefore adversely affecting the client's quality of life.

In this chapter, the core counseling skills and techniques are described, and some relevant models of counseling discussed. However, it is only possible to learn counseling skills through practical application, and it is essential that all practitioners undertake educational courses in counseling skills, at which they are able to practise and develop those skills in a safe training environment. This protects both the client and the practitioner. Such training should include discussion of relevant psychological theory to underpin practice, development of practical competence in counseling skills and exercises that enhance the practitioner's self-awareness.

5.2 The psychological needs of the client and their family

Consideration of the psychological needs of the client is an integral part of the care offered in a genetic healthcare setting. These are inevitably complex and vary as a result of the individual's life changes, as well as the continuous alterations in the family dynamics. The diagnosis or genetic risk for one family member may have serious consequences for relatives. Reactions to the situation that can complicate decision-making or adjustment include:

- guilt at having either passed on or escaped the condition;
- blaming self or other family members for the situation;
- anger.

Being affected or at risk of a genetic condition can cause grief in many forms, and some clients are in a state of perpetual mourning. As many family members may be experiencing difficulties, the usual channels of support through family may not be available. Maintaining secrecy about the condition may also reduce the support available from friends. Genetics counseling may therefore provide an outlet for emotions that are otherwise suppressed. Addressing the psychosocial needs of clients has always been considered a particularly important part of the genetics nurse's or genetic counselor's scope of practice, and is one major area of differentiation between the role of the genetics nurse or counselor and the role of the medical geneticist.

Family systems theory makes a differentiation between first- and second-order change in a family (Dallos and Draper, 2000). First-order change occurs when an externally visible change occurs, but this does not require specific adjustment by the family. For example, when a child whose parents have been convinced he has a familial condition is finally diagnosed by a genetic test. A second-order change requires adjustment by family members, whose relationships and status within the family may alter. This type of change could occur when a family member who has been assumed to have inherited the condition is subsequently found not to carry the mutation, causing others to feel that their own risk has increased. Understanding the need of the family to respond to second-order changes can help the practitioner to support the family through periods of adjustment.

5.3 Person-centered counseling

In every session of genetic counseling, it is important that clients are able to express their own concerns, questions and reactions, and to feel that the genetics practitioner has heard and addressed them appropriately. One model that is suitable for counseling in a genetic counseling context is the person-centered model that is based on the seminal work of Carl Rogers (1961). The central tenet of the model is the belief that each person has the ability to solve his/her own problems and work through difficult situations using one's own resources. Support from another person enables the client to explore the situation in a safe emotional environment.

The aim of person-centred counseling is to facilitate the client to achieve self-actualization through enhancing self-belief. The counselor aims to hold the client in unconditional positive regard, and to demonstrate this. The empathic relationship is central to the counseling work.

Person-centered counseling is appropriate in a genetic healthcare setting, as the practitioner does not profess to be 'an expert', who can solve the client's problems, but rather a supporter whose role is to reinforce the client's self-belief. Rogers described the 'core conditions' that are necessary for a helpful counseling relationship.

Core conditions

Genuineness   The counselor is real to him/her self and to the client. To achieve this, the counselor requires a considerable degree of self-awareness and a belief in the equality of the client.

Empathy   One description of empathy is being able to 'walk in the other person's shoes'. Whereas sympathy involves feeling sorry for the other person, empathy is more connected with trying to understand how the client feels, and communicating that understanding.

Warmth   Understanding the client is not facilitative unless that can be conveyed. The 'gold standard' for the person-centered counselor is the ability to hold every person in unconditional positive regard. This itself is a challenge, but it helps to reduce value judgments of the client and therefore increases the likelihood that the client will feel free to make the decision that is best for them.

The following advanced core conditions should only be used when an atmosphere of trust has been established between the client and counselor.

Immediacy   The use of immediacy in the session helps to reinforce the genuineness of the counselor, and to ground the session in the 'here and now'. Immediacy can also be called 'you–me talk', as the counselor notices and comments on what is actually happening between the client and the counselor during the session.

PRACTITIONER: *Perhaps you don't really want to think about having a termination?*

Client: *I have to, it's not a big deal…*

PRACTITIONER: *I noticed when I said the word termination you looked away. It seems really painful for you to talk about.*

Challenge   Challenge is used when there are discrepancies in the client's account. It is not aggressive, but offered as a way of inviting the client to examine what is happening. For example, the client might say that she had always imagined herself as a mother, then say that it was lucky she didn't have children because she wouldn't have had time for them. This discrepancy in the story could be challenged to help the client become aware of her own difficult feelings.

Concreteness   The counselor uses concreteness to maintain the reality of the discussion and help the client focus on what is currently relevant. For example, a woman seeking predictive testing may repeatedly refer to having the test for her children's sakes. The counselor senses that she is avoiding considering the effect that a bad news result would have on herself, and may ask her to try and exclude the children from her thoughts for a short while to focus on her own feelings.

5.4 Basic counseling skills

The basic tools that the counselor or nurse can use to convey empathic listening are:

- open questions;
- reflection of feelings;
- paraphrases of content;
- summaries of the dialog;
- non-verbal communication;
- silence.

Use of open and closed questions

There are two basic types of question: open and closed. The closed question is one to which the client can reply either yes or no, or reply with a factual answer. Although it is sometimes necessary to use a closed question in a counseling environment, this restricts the client's expression. The use of an open question enables the client to formulate the answer that best fits their agenda and priorities at that time, and is more likely to enable the client to disclose what they are feeling.

Useful forms of open questions include:

- How are you feeling about that?
- Can you tell me more about that?
- I'm wondering how you dealt with that?

Example: Closed question

PRACTITIONER: *Did you feel angry when you were told you were at risk of Huntington disease?*

Client: *Yes, very angry.*

Example: Open question

PRACTITIONER: *How did you feel when you were told you were at risk of Huntington disease?*

Client: *At first, I felt very angry. 'Why did this have to happen to me?' But then I went into a kind of depression. I thought, what's the use of trying, with work, with my girlfriend ....but even at that time I felt guilty for that, like I didn't value my father. I still feel that if I get down.*

Reflection of feelings

One of the most powerful forms of intervention is reflection of the client's feelings. This affirms to the client that he or she has been heard at an emotional level, and can help the client to recognize the veracity and depth of their feelings. The use of the client's own words is particularly potent.

Client: *I just felt ....annoyed, I guess.*

PRACTITIONER: *Annoyed?*

Client: *Yes, well...............more than annoyed................I was quite mad with them for having left it so long to tell me.*

PRACTITIONER: *So, really you were mad?*

Client: *Ummm......I just felt well...it's my life...they didn't have the right to keep that sort of thing from me.......in fact, it was pretty cool between us for a while. I didn't want to see them in case I exploded....*

Paraphrasing content

From time to time during the session, it is helpful to paraphrase the content of the client's story. This confirms to the client that the story has been heard, enables the counselor or nurse to check they have understood and allows the client to correct any misunderstanding. The paraphrase should always be offered tentatively so that the client feels able to correct the practitioner if necessary.

Client: *So I was told by my grandmother that I had a 50/50 chance of getting Huntington disease before I was 30, but if I made it to 30 and I was okay, then I wouldn't have to worry. My father was 30 when he got it. But then, I went to the doctor and he told me that I could get it anytime up till 70, so I was confused and didn't know who to believe.*

PRACTITIONER: *You had conflicting pieces of information about your risk of Huntington disease and don't know which is right?*

Client: *Yeah, I just need to be sure of what I'm dealing with before I make a decision.*

Summaries of the dialog

At key points in the session, and importantly at the end of the session, it is helpful to provide short summaries of the conversation so far. This draws the threads of the dialog together and provides an opportunity to clarify that both parties have understood each other. The final summary might also include plans for action that have been agreed.

PRACTITIONER: *So today we've talked a lot about how you've been feeling since your family told you about your father having Huntington disease. I can see that it's been very difficult for you to deal with, being at risk, and now you're thinking about having a test to see if you really have inherited Huntington's from your father.*

Client: *Yeah, I'm not sure though.*

PRACTITIONER: *What I suggest we do is meet again in a month to talk again about testing. In the meantime, perhaps you can think about the pros and cons and also about who you might tell if you decide to be tested. Would that be okay with you?*

Client: *I guess it seems a long time, but I know I need to think this through properly.*

Non-verbal communication

Because non-verbal communication is mostly unconscious, we are constantly conveying our thoughts to others through our physical actions, without being aware of it. However, it is important in a counseling environment to be aware of the messages that are being conveyed both to and from the client by non-verbal means.

When communication is congruent, the content and feeling of the spoken message 'matches' the non-verbal message. This congruence is important in establishing trust and genuineness. Take, for example, a situation in which a counselor checks her watch while the client is talking:

Response A – incongruence

Client: *Oh, I'm sorry, I'm taking too much of your time.*

PRACTITIONER: *No, no, don't be silly, that's what I'm here for.*

Response B – congruence

Client: *Oh, I'm sorry, I'm taking too much of your time.*

PRACTITIONER: *I'm just checking the time as we have about 10 minutes of this session left, and I want to make sure we have enough time to talk about whether we need to make another appointment.*

Response B demonstrates congruence and will enhance the relationship rather than detract from it.

The health professional must also be observant of the client's 'body language'. It is sometimes appropriate to comment on incongruence between verbal and non-verbal communication. However, this intervention is a type of challenge and should therefore only be used when a degree of trust has been established with the client.

PRACTITIONER: *How did you feel when your girlfriend decided to call things off after she saw your father?*

Client: *I was ...okay about it....I mean, it's her life, her choice....*

PRACTITIONER: *You looked very sad as you said that, I wonder if you really feel okay?*

Client: *Mmmm.....I know it's her choice, I don't blame her a bit.......but I thought she loved me enough to stay and we'd get through it together.*

Silence

One of the most powerful counseling tools is the use of silence. In normal conversation, silence may be viewed as an indication of a lack of social skill. For this reason, the natural inclination is to fill the silence and it is therefore a difficult tool to learn to use. However, silence enables the client and the counselor to think about what is being said, and this often results in moving the discussion to a deeper level. A period of silence after the client has said something highly significant gives weight to what they have said.

The basic counseling skills are discussed in more depth in books listed in the reference section at the end of the chapter, especially Bayne *et al.* (1998) and Heron (2001).

5.5 Non-directiveness

Non-directiveness in counseling has been cited as a gold standard for genetic healthcare. The client is considered to be the best person to make decisions that affect his or her own life, and pressure to act in a certain way is avoided. However, this emphasis on non-directiveness developed partly as a response to the zealous practice of eugenic principles and should be considered in that light, especially as the establishment of genetics as a specialty occurred in the immediate aftermath of World War II, when sensitivity about eugenic practices where extremely high (Kevles, 1999). In practice, it has to be acknowledged that very few interactions in healthcare are completely without some form of directiveness, particularly when health education or health promotion are involved.

The genetics health professional does, however, have a responsibility to the client to enable discussion that assists the client in identifying the most appropriate course of action for that person or family. Although all potential options may be mentioned, the best care for that family may not involve equal discussion of all options, as some of these might not be available to the family or might be inconsistent with their belief systems. Kessler (1997) writes about non-directiveness as meaning a lack of coercion. However, a sensitive practitioner will enable discussion to progress so that the family's preferences are made explicit. The family may benefit from confirmation that their decision is not an inappropriate one in their circumstances. This may be new territory for them, and they might benefit from an indication that their chosen path is reasonable.

Faced with difficult decisions in an unfamiliar situation, some clients will ask the practitioner, 'If you were me, what would you do?'. This is difficult to answer, as in practice it does not seem helpful to decline to respond. One option is to acknowledge the difficulty of their situation empathically and genuinely.

Client: *My sister doesn't know about the Huntington's in our family yet, she's only 16, I don't know if I should tell her.*

PRACTITIONER: *You're not sure if you should tell her yourself?*

Client: *I know how I feel about not being told........I wish they'd told me long ago......but I'm not sure if she'll feel the same way......we're very different.*

PRACTITIONER: *You think she might react differently to you?*

Client: *Yeah....I might make it worse for her..............what do you think I should do?*

PRACTITIONER: *That's such a difficult question for me, too...........there doesn't seem to be a right or wrong answer at this stage....it seems like you really can't know beforehand if it's right to tell her or not.....*

Client: *That's just it....I don't know, but I don't want to keep this secret from her....we've always been so close....she trusts me.....*

PRACTITIONER: *Maybe keeping the trust in your relationship is the most important thing for you both?*

Every practitioner will, of course, have personal beliefs and cultural values that influence their own stance on certain issues. A good professional training scheme will provide encouragement and opportunities for the practitioner to examine their own attitudes on issues that are relevant to their own professional practice. This will enhance insight into the ways in which a client could be influenced. For example, the use of certain vocabulary to describe various options may be indicative of the practitioner's opinion. The power ratio between a client and a healthcare professional is rarely equal, the client perceiving the practitioner as more powerful in a situation that is unfamiliar to the client. Pressure can be exerted in many ways, often subtly. It can be implied simply by the wording of certain choices, or the amount of time spent discussing certain options.

> 'You can choose to terminate the pregnancy or you can keep your baby.'

A client may feel coerced into a decision if the counselor says that most people make a particular decision. For example, 'It is entirely your decision, but in my experience I find most people want to avoid having a child with Down syndrome.'

5.6 Grief, loss, and mourning

Causes of grief

In a genetic healthcare setting, many of the issues faced by clients and their families will revolve around loss. Although this is also true of counseling in many other settings, loss may be repeated many times over many generations in families with inherited conditions. This anticipation and experience of grief becomes part of the fabric of the family. An awareness of the tasks of mourning is necessary for those who are supporting families, whether the loss is a single event, such as a termination of pregnancy for a fetus diagnosed with anencephaly, or a continual pattern of loss in families whose members are at risk of dominant or X-linked conditions.

Although grief is most commonly associated with death, it can accompany any second-order change (see section 5.2). Potential second-order changes are listed in Box 5.1, but this is not intended to be a comprehensive list, as loss is defined by the individual involved. An experience that triggers intense grief in one person may cause no emotional difficulty in another. These differences can be observed between members of the same family, as can differences in the ways that individuals manage the tasks of mourning. These can be the cause of misunderstanding in the family and one of the ways in which support can be offered is to emphasize the 'normality' of the reactions of each person to their situation and enhance the family's understanding of the processes involved.

Box 5.1 Loss situations that may be experienced by family members in a genetic healthcare setting

- Death of family member
- Diagnosis of condition in self
- Diagnosis of condition in family member
- Loss of health
- Disability
- Loss of personal independence
- Loss of pregnancy
- Loss of 'normal' baby or child
- Loss of partner due to their changing role in family
- Loss of personal freedom due to commitment to caring
- Loss of ability to drive
- Loss of ability to earn a living
- Loss of financial security
- Awareness of risk to self
- Awareness of risk to family members
- Loss of normality

In addition to the immediate loss, a feeling of ongoing grief may result. For example, one couple who lost two children in infancy described not only the loss of the children, but their continual hurt at seeing their friends going through the tasks and rituals of parenthood. Even everyday occurrences, such as seeing their friends walk their children to school, could be painful. However, as those children graduated, married and had their own children, the loss of the joys of parenthood and grandparenthood were continually impressed on them (Skirton *et al.*, 2004).

The tasks of mourning

The term 'mourning' describes the work that must be done to resolve grief. Building on the work of previous authors such as Kubler-Ross (1969) and Worden (1991) identifies four main tasks of mourning. These are not achieved in a linear sense – that is, it is not necessary or perhaps even possible to complete one before moving to the next, but each step needs to be addressed at some level before the next can be achieved satisfactorily. The experience of most people is that they oscillate between tasks. There is no 'normal' time for mourning – the time needed by each individual should be respected, but many people who have lost a close family member (such as a partner or child) find that they are unable to invest emotionally in the future for a period of 2 to 5 years following the loss. Clients may be concerned about their own mental health and sometimes benefit from the reassurance that what they are experiencing is not abnormal.

Acceptance of the loss   The initial response to loss is, frequently, shock and denial. This may be momentary or may last for an indefinite period of time. There may be more difficulty in accepting the loss when it is entirely unexpected, but even expected news may be hard to absorb initially. Clients often describe feeling 'numb' during this stage.

A mother called the genetics nurse three weeks after the death of her son, Daniel, who had been diagnosed with Duchenne muscular dystrophy when he was 2 years old. Daniel was 15 years old at the time of his death and had been wheelchair bound for 6 years. He had been treated in intensive care for pneumonia during the previous year. After that episode, Daniel told his family that if he became acutely ill again he did not want intensive treatment. When he contracted pneumonia again, he was taken to the local hospital and his wishes not to have intensive therapy were respected. His mother told the nurse that she could not believe he was dead and was constantly expecting him to come home. She felt guilty because she was feeling numb and unable to cry.

Feeling the pain of the loss   Following at least partial acceptance of the loss, the mourner will experience the painful feelings of grief, which are characterized by periods of uncontrollable weeping or anger. These may be so intense that the mourner is unable to perform the normal functions of everyday life, or may feel that they are just 'going through the motions'. Grief is a very draining experience, and at such times the client may need to be given 'permission' to perform only those tasks that are necessary for survival.

Harry and Jane were shattered when their father Ron died from an aortic anuerysm when he was 46 years old. Ron had been diagnosed with Marfan syndrome at post-mortem. His children were angry that their mother Helen did not seem to wish to support them emotionally. In fact, she seemed to be avoiding them. Helen had always 'been there for them', but the grief of Ron's loss was so intense, she had no energy to support anyone else and found conversations with her children almost unbearable.

Adjusting to the loss   During this stage, the grieving person gradually adjusts to life with a different emotional and physical landscape. The bereaved person accepts that the loss is not reversible and makes the necessary changes to integrate the loss into their everyday life.

Martin was diagnosed with autosomal dominant retinitis pigmentosa when he was 27 years old. At first, his vision was adequate to enable him to continue driving, but when he reached 30 years old, he was told his field of vision was so impaired he could no longer legally hold a license to drive. He rebelled at first, refusing to go out. Gradually Martin was able to integrate the loss of his ability to drive into his life and accepted his wife could drive when they went out.

Investing in the future   If the tasks of mourning are managed adequately, the individual will be able to focus increasingly on the future rather than on the loss.

Jack and Kate had a termination of their first pregnancy after anencephaly was diagnosed at 16 weeks gestation. They both found it difficult to accept the loss and were afraid of a similar problem occurring again. After a year, Kate felt ready to try again for another pregnancy, but Jack was still very distressed and did not even want to consider it. While Kate had the support of friends and her sister after the loss of the baby, Jack had been putting on a brave face to support her. When Jack realized she was feeling stronger, he allowed himself to grieve for the baby and about a year later felt he could look to the future and have another pregnancy with some confidence.

In these cases, the practitioner can facilitate the individual in managing the tasks of mourning by:

- empathic listening;
- identifying the situation as a loss;
- emphasizing the normality of the person's reaction;
- recognizing if grief is complicated and making an appropriate referral to another health practitioner.

Complicated grief is defined as grief that does not resolve itself and that is seriously impeding the ability of the mourner to invest in their changed life. However, the time required by those who have suffered a loss varies greatly and making a diagnosis of complicated grief may require the input of a trained counselor.

5.7 Using a transactional analysis (TA) model

One counseling model that can be used by the practitioner to aid understanding of the client's behavior is the TA model. This was developed by Eric Berne (1991) and is based on the belief that all humans are born believing in their own worth and the worth of others, but that this belief often becomes damaged during childhood. The belief in your own self-worth is called the 'I'm okay' position, whereas belief in others is termed 'You're okay'.

The aim of counseling is to enhance the client's belief in themselves, helping them to return to the 'I'm okay, you're okay' state that is necessary for healthy relating. Another aspect of the TA model is the use of the terminology 'parent', 'adult' and 'child' ego states. The ego state mediates between the person's inner world and the external world, in effect controlling their behavior. The three ego states exist in each person, one being dominant at any particular moment in time. The parent ego state may manifest as the critical or controlling parent, which reinforces duty messages such as 'I should'. The protective parent contains the nurturing aspects of the ego state and may be induced in situations in which the other person is seen as vulnerable. The free child is the more natural, spontaneous ego state, whereas the adapted child responds in a way that would bring approval from others. The ego state in which the relative aspects of each course of action are considered logically is called the adult state.

TA can help to explain why we sometimes react seemingly irrationally with others. For example, a person whom we find very dominant can evoke a reaction from our child state. The counselor may try to facilitate the client in responding from the adult state as far as possible. The TA counselor works to help the individual achieve the 'I'm okay, you're okay' position. This may be highly relevant in families affected by genetic conditions, particularly where the style and quality of parenting has been affected by the presence of the disease.

5.8 Transference and counter-transference in a genetic counseling setting

During any personal interaction, there is the potential for one person, such as a client, to 'transfer' feelings about another onto the other person (such as a counselor) who is involved in the interaction (Jacobs, 1985). This commonly occurs when some aspect of the counselor 'reminds' the client of another person with whom they have had emotional contact. In psychotherapeutic counseling, this transference is used as part of the therapeutic process, with the counselor often representing the 'good enough' parent to the client, enabling the client to grow emotionally in a 'secure enough' environment.

In genetic healthcare interactions, the practitioner may be perceived to be in a powerful position as 'expert'. Transference of feelings that have arisen when the client has been in situations that are emotionally similar may cause the client to respond in ways that are inexplicable to the practitioner.

For example, if the client had a very dominant and critical mother, the experience of being with a female expert practitioner may stimulate feelings of fear or submissiveness in the client.

Counter-transference can also occur, with the practitioner responding emotionally to the client in ways that are not explicable in the context of the professional relationship. Counseling training and supervision aim to make the practitioner aware of both transference and counter-transference for the safety of both clients and practitioners.

CASE STUDY – KALIA

Oona was a practitioner working in a cancer genetics clinic. As part of her role, she took a medical history from a woman called Kalia who had a strong family history of colorectal cancer. Familial adenomatous polyposis was subsequently confirmed in the family. At a clinical appointment to discuss the risks to the children in the family, Kalia became very angry. After some discussion, it emerged that she was very angry with her mother for not telling her earlier about the family history and enabling her to be screened. Oona's understanding of transference enabled her to continue to support Kalia emotionally rather than reacting to her anger.

5.9 Counseling supervision

It is imperative that practitioners who provide counseling support for clients as part of their genetic healthcare should use a counseling supervisor (Bayne et al., 1998). The aim of supervision is to protect both the client and counselor from emotional harm and to maximize the benefit of the counseling relationship to the client. The role of the supervisor is to enable the practitioner to explore the meanings of their relationships with clients, to discuss both previous counseling sessions and to prepare for future sessions. Personal issues belonging to the counselor will inevitably sometimes impinge on their work, and the supervisor will encourage the counselor to explore and reflect on these. The supervisor is normally an experienced counseling practitioner (not necessarily in genetics) who has undertaken additional training in supervision skills.

Group supervision may sometimes be arranged for a team of practitioners. This has the benefit of enabling practitioners to learn from the experience of others, but may inhibit open discussion if there are differences in the power that is held between practitioners (for example, some are experienced and others are novices) or if the group is not facilitated in a way that makes disclosure safe for everyone.

5.10 Conclusion

Families who are referred for genetic healthcare may have psychological needs connected with the genetic condition or risk in their family. These may be complex and are often related to loss. Developing appropriate counseling skills is an essential part of the preparation for competent practice of genetic healthcare practitioners. Although experience in working with families aids the development of communication skills, formal counseling training enables the practitioner to practise skills in a safe setting and to enhance their own self-awareness. The use of counseling supervision individually or in a group helps to protect the safety of both practitioner and client.

Study questions

1. Case scenario – Eva and Juan

Eva and Juan are a couple in their mid-40s. They have been married for 15 years.

Eva has had four confirmed pregnancies. In each pregnancy, she has miscarried at between 6 and 9 weeks gestation. After the third miscarriage, both partners had chromosome studies and Juan has a balanced chromosome translocation.

Task A
Suggest the issues of loss that might be significant for this couple.

Task B
Discuss how their responses to Juan's karyotype results might differ.

Task C
Eva calls to say she is 4 weeks pregnant. As a practitioner, how might you provide psychological support to this couple?

Task D
You and your partner have recently lost a pregnancy, how might this affect your care of this family? What action might you take?

2. Case scenario – Milly and Karen

Milly and Karen are a couple in their mid-30s who have been living together for 2 years. Milly has told Karen about her family history of colon cancer, but said she didn't expect to get it because she was 'nothing like her father'.

At a routine medical before starting a new job, Milly is referred for genetic counseling with regard to her family history.

She is actually at a 50% risk of inheriting a mutation for hereditary non-polyposis colon cancer and screening by colonoscopy is recommended.

Task A
Suggest the potential second-order changes for Milly. Do the same for Karen.

Task B
Milly refuses to have a colonoscopy because she is sure she won't get the cancer. How might the use of Roger's core conditions help you to address this with her?

References

AGNC (2004) Competencies for UK Genetic counselors. [Accessed March 4, 2004] www.agnc.co.uk/Registration/competencies

American Society of Human Genetics Ad Hoc Committee on Genetic Counseling (1975) Genetic Counseling. *Am. J. Human Genet.* **27**: 240–242.

Bayne, R., Nicolson, P. and Horton, I. (1998) *Counselling and Communication Skills for Medical and Health Practitioners.* The British Psychological Society, Leicester.

Berne, E. (1991) *Transactional Analysis in Psychotherapy.* Souvenir Press, London.

Dallos, R. and Draper, R. (2000) *Introduction to Family Therapy.* Open University Press, Maidenhead.

Heron, J. (2001) *Helping the Client.* Sage Publications, London.

Jacobs, M. (1985) *The Presenting Past.* Open University Press, Milton Keynes.

Kessler, S. (1997) Psychological aspects of genetic counseling. XI. Nondirectiveness revisited. *Am. J. Med. Genet.* **72**: 164–171.

Kevles, D.J. (1999) Eugenics and human rights. *BMJ* **319**: 435–438.

Kubler-Ross, E. (1969) *On Death and Dying.* Macmillan Publishers, New York.

Rogers, C.R. (1961) *On Becoming a Person.* Constable, London.

Skirton, H., Evans, E. and Evans, R. (2004) More than just science – one family's story of the diagnosis of a chromosome translocation. *Paediatr. Nurs.* **16**(10): 18–21.

Worden, J.W. (1991) *Grief Counselling and Grief Therapy,* Second Edition. Routledge, London.

Further resources

Burnard, P. and Hulatt, I. (1996) *Nurses Counselling.* Butterworth-Heinemann, Oxford.

Cozens, J. (1991) *OK2 Talk Feelings.* BBC Books, London.

Egan, G. (1998) *The Skilled Helper.* 6th Edn. Brooks/Cole Publishing Company, Pacific Grove, CA.

Frankland, A. and Sanders, P. (1995) *Next Steps in Counselling.* PCCS Books, Manchester.

Hough, M. (2000) *A Practical Approach to Counselling.* Longman, Harlow.

Jacobs, M. (1985) *The Presenting Past.* Open University Press, Milton Keynes.

Mearns, D. and Thorne, B. (1999) *Person-Centred Counselling in Action,* Second Edition. Sage Publications, London.

Murray Thomas, R. (1990) *Counseling and Life-Span Development.* Sage Publications, Newbury Park.

6 Working as an educator for families and professionals

6.1 Introduction

The genetics health professional works as an educator in a number of different ways:

- with patients and their families to assist in managing disease, and understanding and using genetic information;
- with other health professionals to educate and inform them about genetics and its relation to health and disease;
- with students;
- with colleagues;
- with the wider public.

Any educational activity is an interaction between two participants, the educator and the learner. For the genetics health professional, most of their work will be in the context of working with adults either outside of a formal educational setting or in the context of professional education. For that reason, it is helpful to consider some of the principles that underpin educational theories of adult learning. There is a considerable body of literature in the field of education and this chapter can provide no more than an overview of some of the schools of thought in this area. However, those genetic healthcare practitioners whose role involves substantial teaching may wish to use the 'Further resources' to read more widely.

6.2 Expressing and understanding risk

Clinical genetics is an area of medical practice in which, traditionally, much focus has been placed on transmitting information about risk. However, it should be emphasized that, with the explosion of medical information now available and the shift from paternalistic to shared decision-making in the clinical encounter, how risk is communicated is relevant to all healthcare.

Normally, the purpose of communicating risk is to enable some kind of decision to be made. If the purpose of the clinical encounter is to facilitate the decision-making, then presenting the facts about risk should be more than simple information giving. For reviews on genetic counseling and decision-making, see Michie and Marteau, 1996; and Shiloh, 1996. The challenge is to turn the numerical risk information into meaningful information that can be used by the client.

The language of risk

Terms such as 'probably', 'likely', 'low risk', etc. have different meanings for different people. One person may regard a risk of 1% having a baby with Down syndrome as low, whereas another may regard it as high. Many genetic health professionals have been in a clinical situation in which they are informing someone that they are at a 50% risk of inheriting a condition such as Huntington disease and have been met with expressions of relief because the client thought they were at a 100% risk and now its 'only' 50%.

Understanding of numerical risk is also complicated by the different ways in which the number can be presented. For example, in cystic fibrosis the recurrence risk after the first affected child is born may be presented in a number of different ways:

- 1 in 4 that a future child will be affected;
- 3 out of 4 that a child will not be affected;
- 25% risk of having a baby with cystic fibrosis;
- 75% chance that the baby will not have cystic fibrosis;
- odds of 3 to 1 that the child will be unaffected;
- odds of 1 to 3 against the baby being unaffected;
- odds of 1 to 3 that the baby will be affected.

In the above examples, the way in which the numerical risk is presented has been altered as well as the type of language that has been used. These changes in language are called 'framing effects' and it is well-recognized that decisions will be affected by the way in which the information is presented.

In decisions about accepting screening, the factors that have been shown to increase uptake of screening are presenting risks as relative risks rather than as absolute risks; presenting the losses from not having screening rather than the potential gains; presenting the chance of positive outcomes rather than of negative outcomes and providing less complex information (Edwards *et al.*, 2002).

The general principles for presenting risk are that information should be presented in a balanced manner with different numerical formats of the same risk. Risk information should be relevant to the individual and should be presented clearly, avoiding information overload.

Understanding risk

There is a considerable body of evidence that people do not solely base their genetic decisions on a numerical recurrence risk value (Halbert, 2004). In addition, the individual's perception of the burden of the disease has a huge impact on their decision, and this is a very personal construct based on their own experience of the condition, what they have read or been told, and experience of similar situations (Skirton, 2001).

'Lay knowledge' or 'lay belief' is a term used to describe the information that exists in the family about the particular genetic condition that affects them (Richards and Ponder, 1996). The beliefs are usually rooted in the family's experience of the condition, but may also be based on general health beliefs, or superstition or medical information that has been previously passed on to them. Attributes that are unconnected to the genetic condition (such as hair color) may be linked to the experience of the disease, leading people in the family to believe, for example, that all the redheads in the family will get the condition. In other cases, the disease is linked to the sex of those affected and, if the only affected members of a family have all been male, the family may believe that only sons in the family are at risk.

As many people have only a sketchy idea of genetics, they may seek non-genetic explanations for what has happened in the family. Other information about influences on health may be used to draw conclusions about the cause of a disease or syndrome. For example, a mother who has a child with learning delay may question the quality of her diet during the pregnancy.

Application of lay knowledge will often lead to particular family members being identified as either being at high risk of the condition or likely to avoid it. This pre-selection of affected members can liberate those who do not fit the criteria from worry, but may place extra burdens on those predicted to be affected.

For genetic information to be useful for families, they must be able to fit any new information into their family story. It is helpful to use the family's own family tree when explaining the inheritance pattern, so that any discrepancies between the explanation and the lay knowledge held by that family can be addressed. If this is not done, the lay beliefs will persist, as the family's own experience is more powerful than an abstract scientific explanation. Family dynamics can be disrupted if the lay knowledge is challenged by new developments. The family may require time and support to adjust to a new way of thinking about the inheritance of the condition.

In making decisions, the family will put weight on the burden of the disease as well as the chance of a family member being affected. If the burden of the disease is heavy, even a very low risk may seem oppressive. On the other hand, if the family views a disease as mild, they may consider any risk worth taking.

6.3 Principles of adult learning

As early as the 1920s writers were suggesting that there were differences in adult learning, which meant that the approach to education for adults may need to be framed in a different way from that used to teach children. It may be useful to the genetic healthcare practitioner to be aware that there is a distinction between compulsory education, where the student is there because they have to be (as is the case for most children and many students), and education that is sought either for personal or professional development. A contrast has been made between what was known as a 'pedagogical' style of teaching that was current in teaching children, and an alternative style of teaching involving an interaction based on partnership between the educator and the learner (andragogy).

Lindeman, writing in 1926 (cited in Knowles *et al.*, 1998), identified several key assumptions about adult learners, which have formed the basis of adult learning theory. These are:

- adults are motivated to learn as they experience needs and interests that learning will satisfy; therefore, these are the appropriate starting points for organizing adult learning activities;
- adults' orientation to learning is life-centered; therefore, the appropriate units for organizing adult learning are life situations, not subjects;
- experience is the richest resource for adults' learning; therefore, the

core methodology for adult education is the analysis of experience;

- adults have a deep need to be self-directing; therefore, the role of the teacher is to engage in the process of mutual enquiry with them rather than transmit his or her knowledge to them and then evaluate their conformity to it;
- individual differences among people increase with age; therefore, adult education must make optimal provision for differences in style, time, place, and pace of learning.

During the following decades, developments in psychological thinking that influenced the development of counseling – by people such as Carl Rogers, Erik Erikson, Maslow, and the psychodynamic theorists – contributed to the development of theories of adult learning. As discussed in Chapter 5, a person-centered humanistic framework underpins the practice of genetic counseling and this is mirrored in this developing view of adult education.

In any framework for educating the adult learner, the core adult learning principles interact with and are informed by the goals and purposes for learning, and individual and situational differences.

In this schema, the goals and purposes for learning relate both to the goals of the individuals and those of institutional and societal goals. Individual and situational differences relate to education in practice and shape the educational intervention. The core principles of adult learning are:

- the learner's need to know;
- self concept of the learner;
- prior experience of the learner;
- readiness to learn;
- orientation to learning;
- motivation to learn.

These principles focus directly on the learner.

6.4 Learning theories

There is a distinction between the type of learning that is desired (acquisition of new knowledge, or the development of skills, for example) and the process by which that learning happens. As has been discussed previously, learning is an active process and is a positive action on the part of the learner.

There are numerous theories of learning but they may be broadly categorized into three groups.

Behaviorist theories

This group of theories emphasize the role of the teacher as the agent in learning. The teacher provides a stimulus for learning and also the appropriate responses (rewards and punishments) to reinforce learning.

Cognitive theories

These theories focus on the learner as the active processor of knowledge. However, there is an assumption that there is a body of knowledge that is framed by the teacher and that the learner seeks to organize and master. Both behavioral and cognitive theorists suggest that there are hierarchies of learning, which are sometimes categorized as surface and deep (Ramsden, 1992). The tasks of learning, as framed by cognitive theorists, range from simple signal learning through knowledge building with words to knowledge building with concepts and ending with complex problem-solving (Gagne, 1985).

Humanist theories

This group of theories emphasize the active process of learning as a product of the learner and the social setting. The principles of adult learning presented earlier are an example.

There are some common elements emerging that relate to the transfer of knowledge versus the construction of knowledge, and mechanical learning versus creative learning. Education for adults will tend to focus on creative learning, resulting in the construction of knowledge, which is relevant for the individual's needs at that time. The role of the educator is to facilitate that process.

Experiential learning

Within education, particularly professional education, there is a growing recognition that experience forms the basis of all learning. This is stressed by Mezirow who suggests that the heart of all learning is the search for meaning in experience. Learning is creating meanings and making sense of experience (Mezirow, 1981).

The process by which that meaning is constructed is through critical reflection on experience. It is suggested that there is a learning cycle – starting with experience, proceeding through reflection, and leading to action, which then becomes the concrete experience for further reflection and so on (Schon, 1983).

However, a focus on experiential learning as the primary way of learning tends to underestimate the importance of learning as an activity to acquire new knowledge.

Learning styles

It is suggested that each of the stages of the learning cycle calls for different learning approaches that will appeal to different individuals as we each have a preferred learning style. The learning styles are:

- active learners who prefer to get on quickly with a task and learn by doing something immediately;
- reflective learners who prefer to think things through and reflect on a situation;
- theorizing learners who prefer to understand general principles and concepts rather than specific cases;
- experimental learners who want to try things out for themselves.

Practical applications

In the clinical genetics scenario, the health professional in his/her role as an educator is responding to the needs of the client or the student. Both of these groups – but particularly the client – will be motivated to learn because of their own personal and social circumstances.

Figure 6.1 Learning cycle

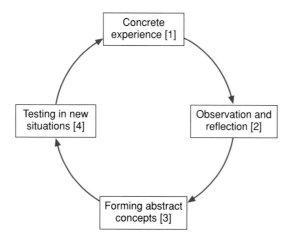

The clients and students will already have their own experiences of learning, will have developed their own preferred ways of engaging in the learning process, and will have particular goals for the outcome of the learning situation. The role of the educator is to use and enhance the participants' existing learning techniques and to make them more efficient, to help learners to move from specific issues to a deeper understanding of their area of concern, to help make the learning permanent and available for later use, and to provide a framework for continuing learning.

The characteristics of what might be called learning 'episodes,' and the implications for the educator are presented in Table 6.1.

The encounter should be used to motivate the student for further learning and to encourage an interest in general principles rather than situation-specific knowledge (Rogers, 1998).

Table 6.1 Learning episodes and their implications

Characteristic	Implications
Episodic	Short bursts of learning material Break material into small units but related to other items of learning
Problem- or situation-specific	Make relevant to the client's/student's needs Be aware of the client's/student's motivations Clients and students to set own goals Start from where the clients/students are
Learning styles	Be aware of different learning styles, which will need different learning strategies
Based on existing knowledge and experience	Relate new material to existing knowledge and experience
Trial and error	Experiential learning Need for feedback Need for practice
Understanding of underlying concepts	Move from simplified concepts to more complex Help students/clients to build up units to create the concept, selecting essential from non-essential units
Imitation rather than memory	Rely on understanding for retention not memory Use demonstration

6.5 Use of internet resources

The success of the sort of learning for all participants in the approaches discussed earlier depends on access to high-quality information. This is true of all healthcare.

The internet has developed into a powerful tool that is used by patients and professionals as a source for health information. The evidence-based medicine movement has focused attention on the quality of evidence that is used to make clinical decisions. Although there is debate about the value judgments that have been placed on certain types of evidence, it has encouraged a more systematic and rigorous identification and appraisal of the mass of information that is now available (Sackett *et al.*, 2000).

The internet has also provided a powerful tool for patients (Taylor *et al.*, 2001). There has been concern about the quality of information, but it is suggested that the harms have been overestimated and the benefits underestimated (Eysenbach, 2003). In addition to being a source of knowledge, patient support groups have provided an important source of online support for nearly all medical conditions.

In discussing education, emphasis was put on educational activity as a partnership between professional and patient. The resource of the internet is a powerful tool for that partnership in empowering and informing the professional and the 'expert' patient (Fox and Lee, 2000; Muir Gray, 2004). Genetics professionals may be able to guide families to evaluate the source of material on the internet and to recognize indicators of information that come from reputable sources, are current and do not present a specific point of view (Snow, 2001).

The internet is now a vital tool for professionals and patients and all genetics healthcare professionals should be aware of how to use it effectively.

6.6 Conclusion

In this chapter, we have talked about the practitioner as educator. There are overlaps between practice as an educator and practice as a counselor. Both require an understanding of the student or client, an ability to start from where they are and facilitate their journey towards their individual goals. The practitioner has the responsibility to equip themselves with the tools and skills to do this successfully.

Study questions

Case scenario – Pippa

Pippa, a genetic counselor in training, has recently joined the genetics department to gain experience in pediatrics. Her previous experience was in adult medicine. She is contacted by a nurse to talk about a patient who has recently been diagnosed with Hunter syndrome. The pediatric unit nurse would like Pippa to come and do a seminar for them. Pippa asks you as a more experienced colleague if you would do it. You think that it would be good experience for Pippa and say you will support her in preparing and giving the seminar.

Task A
Think through how you might help Pippa structure the seminar and decide what to present.

Task B
Using the internet as a resource, find out about Hunter syndrome. Make notes on clinical features, diagnoses and treatments.

Task C
Prepare a plan for the seminar, including learning objectives for the ward nurses.

Task D
After a successful seminar, Pippa is asked to see the family. Think about how you might help Pippa prepare for the family. Write a plan for what you might cover with Pippa in a learning session before she sees the family.

References

Edwards, A., Elwyn, G. and Mulley, A. (2002) Explaining risks: turning numerical data into meaningful pictures. *BMJ* **324**: 827–830.

Eysenbach, G. (2003) The impact of the internet on cancer outcomes. *CA Cancer J. Clin.* **53**: 356–371.

Fox, S. and Lee, R. The online health care revolution: how the web helps Americans take better care of themselves. (Pew Internet & American Life Project). *BMJ* [Accessed March 3, 2005] www.pewinternet.org/reports/toc.asp?Report=26

Gagne, R. (1985) *The Conditions for Learning.* Holt, Rinehart & Winston, New York.

Halbert, C.H. (2004) Decisions and outcomes of genetic testing for inherited breast cancer risk. *Ann. Oncol.* **15** Suppl 1:I35–I39.

Knowles, M.S., Holton, E. and Swanson, R. (1998) *The Adult Learner,* Fifth Edition. Butterworth Heinemann, Woburn.

Mezirow, J. (1981) A critical theory of adult learning and education, *Adult Educ.* **32**: 3–24.

Michie, S. and Marteau, T. (1996) Genetic counselling; some issues of theory and practice. In: *The Troubled Helix: Social and Psychological Implications of the New Genetics.* (eds T. Marteau and M. Richards). Cambridge University Press, Cambridge, pp. 104–122.

Muir Gray, J.A. (2002) *The Resourceful Patient.* Rosetta Press, Oxford.

Ramsden, P. (1992) *Learning to Teach in Higher Education.* Routledge, London.

Richards, M. and Ponder, M. (1996) Lay understanding of genetics: a test of a hypothesis. *J. Med. Genet.* **33**: 1032–1036.

Rogers, A. (1998) *Teaching Adults.* 2nd Edn. Open University Press, Buckingham.

Sackett, D.L., Strauss, S.E., Richardson, W.S., Rosenberg, W. and Haynes, R.B. (2000) *Evidence Based Medicine: How to Practice and Teach EBM.* Churchill Livingstone, Edinburgh.

Schon, D.A. (1983) *The Reflective Practitioner: How Professionals Think in Action.* Basic Books, New York.

Shiloh, S. (1996) Decision making in the context of genetic risk.I In: *The Troubled Helix: Social and Psychological Implications of the New Genetics.* (eds T. Marteau and M. Richards) Cambridge University Press, Cambridge, pp. 82–103.

Skirton, H. (2001) The client's perspective of genetic counselling – a grounded theory approach. *J. Gen. Counsel.* **10**: 311–330.

Snow, K. (2001) The growing impact of genetics on health care: do we have appropiate educational resources? *Mayo Clin. Proc.* **76**: 769–771.

Taylor, M.R.G., Alman, A. and Manchester, D.K. (2001) Use of the internet by patients and their families to obtain genetics-related information. *Mayo Clin. Proc.* **76**: 772–776.

Further resources

Foundation for Blood Research Southern Maine Genetics (2003) *A Genetics Resource Guide for Nurses.* Order via: www.fbr.org

Knowles, M.S., Holton, E. and Swanson, R. (1998) *The Adult Learner,* Fifth Edition. Butterworth Heinemann, Woburn.

Ramsden, P. (1992) *Learning to Teach in Higher Education.* Routledge, London.

Rogers, A. (1998) *Teaching Adults.* 2nd Edn. Open University Press, Buckingham.

Schon, D.A. (1983) *The Reflective Practitioner: How Professionals Think in Action.* Basic Books, New York.

7 Working as a researcher

7.1 Ways in which research contributes to genetic counseling

The body of knowledge in genetic healthcare is drawn from a variety of sources, including research, clinical experience, expert opinion, and tradition. As the scientific basis for genetic healthcare has grown and matured, a greater proportion of knowledge that guides practice reflects the results of systematic inquiry. One way to define research is that it is a systematic process of inquiry to validate and refine existing knowledge, and to generate new knowledge (Burns and Grove, 2001). When applied to genetic healthcare, this research focuses on situations that include genetic assessment, education, counseling, and genetic aspects of health and disease in individuals, families, populations, and communities. Much research in genetic healthcare could be considered to be clinical research, in which findings from basic research – for example, molecular genetics or behavioral sciences – are applied to clinical problems. Both the International Society of Nurses in Genetics and the Association of Genetic Nurses and Counsellors endorse the participation of healthcare professionals in the genetics research process (ISONG, 2004; Skirton et al., 1998). Genetics practitioners may participate in the generation of new knowledge through research in all aspects of the research process.

7.2 The research methods continuum

Clinical research can be divided into two categories: quantitative and qualitative research (Table 7.1). The category referred to as quantitative research is the traditional research process in which information gained from one study is not regarded as sufficient for inclusion into the body of science until it has been replicated several times with similar findings each

Table 7.1 Comparison of quantitative and qualitative research approaches

Quantitative	Qualitative
Used in basic and applied research	Used to generate new insights, increase sensitivity, identify new phenomenon
Deductive	Inductive
Rigor and control in study design	Adhere to philosophic perspective in study design
Sample size reflects power analysis	Sample size reflects point of saturation
Control in sampling and setting to reduce extraneous variables	Diverse sample; context contributes to understanding of data
Objective measurement. Administered consistently throughout study	Data from observation or via interview; may change as study progresses
Data analyzed at completion of data collection	Data analyzed during data collection
Statistical analysis to determine probability that results are accurate reflection of reality	Analysis of data seeks common themes; explanation of social processes
Study can be replicated	Analysis not replicable
Results may be generalizable	Results not generalizable

time. This approach to research is a formal objective process that is designed to describe and test relationships, or to examine cause-and-effect interactions among variables (Burns and Grove, 2001).

Assumptions in quantitative research

Several assumptions are the foundation of quantitative research. One of these is that truth is discoverable, but that the discovery is always incomplete or imperfect. Therefore, hypotheses or study predictions are not proven, but they are either supported or not supported by the findings of the study. Quantitative studies are characterized by rigor and control, meaning that the study is carefully designed and implemented to decrease the possibility of error and to increase the probability that the findings are an accurate reflection of reality (Burns and Grove, 2001). Specific definitions of the population and variables, and consistency in the administration of a treatment and measurement of variables, all reflect efforts by the researcher to achieve a high level of control in quantitative research projects. Random sampling and use of standardized measures are two components of quantitative research, which contribute to the intent that results can be generalizable to populations beyond the sample that was obtained for that specific study. Historically, most healthcare research has been in the quantitative category.

An example of quantitative research in genetic healthcare is the investigation of knowledge and emotional consequences of cystic fibrosis (CF) population screening (Gordon *et al.*, 2003). In this study, individuals who completed CF carrier testing completed a paper and pencil instrument that documented knowledge, and another that measured emotional consequences. The mean scores were compared between those who were carriers and those who were non-carriers. Use of this method could allow replication of the study with other clinical populations.

Different philosophic assumptions form the basis for qualitative research methods. There are several types of qualitative research approaches, and each of these has specific language and procedural rules that determine the rigor and scholarliness of the research. Some examples are the use of qualitative methods to conduct a descriptive study, grounded theory methods to develop a theoretical explanation for a social process, or ethnographic methods to examine research questions in a particular culture (Burns and Grove, 2001). In all instances, qualitative methods are used when little is known about a phenomenon, when the researcher wants to increase the sensitivity of healthcare providers to a clinical topic, or when data are needed to develop a new research instrument (Knafl and Webster, 1988).

Assumptions in qualitative research

Qualitative research methods share the assumption that there is not a single reality, and that the context contributes to understanding of truth. In qualitative studies, there is no attempt to control the research environment; data are elicited through interviews or other methods of expression of the concept of interest, such as observation of participants in their natural setting, journaling, or photographs. The study of perceptions of persons serving as support persons throughout genetic counseling for predictive Huntington disease (HD) testing (Williams *et al.*, 2000) is an example of a qualitative research study in which concerns of persons serving as support persons were described. In this study, sampling was purposive, data were collected through semi-structured interviews, and questions were revised as new insights were gained during the course of the study. Although the results of this study are not generalizable to other populations, the results are useful in generating insights and questions that could be explored through quantitative methods, or through clinical observations. Some research questions are most appropriately answered through traditional quantitative research methods, whereas others are not, and require a qualitative approach. An organized approach to reading research reports is useful in evaluating the study's strengths and weaknesses (Box 7.1).

Box 7.1. Critical analysis of research reports

- Is the problem significant?
- Is the reasoning behind the study logical?
- Is the literature review complete and current?
- Is the design appropriate to answer the research question/or to test the hypotheses?
- Are measures appropriate for the study sample?
- Do measures have appropriate reliability and validity?
- Are there potential biases in the sampling that may influence the findings?
- Is the data collection process appropriate and consistent throughout the study?
- Is the analysis clear and consistent with the level of data obtained?
- Are the conclusions consistent with the findings?

Burns and Grove, 2001

7.3 Elements of the research process

Genetic healthcare professionals who have clinical responsibilities are in an excellent position to recognize researchable problems. Frequently, the impetus for a research project starts with an astute clinician who says, 'I wonder.' This was the situation when a practitioner wondered if genetic testing of offspring of carriers of balanced translocations was beneficial or harmful to the children. In this study, Barnes (1998) reported that the majority of children had not been informed of their translocation, but in those who had been told, the parents reported no negative reactions to the news. The method she used was a descriptive quantitative research design in which adult carriers of balanced translocations responded to a mailed questionnaire. The results of this study revealed insights into the challenges faced by parents who pondered when and how to inform their children.

Regardless of the research methods used in clinical studies, there are several components that all clinical research studies have in common. One is that the research addresses an important healthcare problem. A research problem is a situation in need of description, explanation or improvement. The research problem tends to be a very broad statement, such as what is the disclosure experience of balanced translocation carriers, but no single study can answer all research questions. Therefore, many studies, with different research purposes and aims, may be developed from the identification of one research problem. However, all research studies will have a specific type of design that is most appropriate for fulfilling the purpose of the study. For example, Berhnhardt and colleagues (2003) designed a study with the specific purpose of assessing parent and child reactions to participation in disease-susceptibility research. The study used a descriptive qualitative method in which parents and children participated in audio-taped interviews. The researchers found that a child's perception of risk was not consistent with the perceptions of the parents, or that of the healthcare providers. The results of this type of study could then be used to develop and test hypotheses, or interventions in studies that employed quantitative methods.

After the research problem is recognized, or identified, often a team of researchers develops a research plan, seeks funding and implements the research. The next portion of the chapter discusses the various types of research that may be used.

7.4 Relevant research methods: qualitative

Qualitative research is used when there is little known about a problem, when there is a desire to increase sensitivity of healthcare providers, or when data are needed to develop new measures. Skirton's (2001) study of the expectations of individuals who seek genetic counseling is an example of the first reason to conduct a qualitative study. In this study, she used a grounded theory methodology and purposive sampling to identify individuals seeking genetic counseling. Data were collected through semi-structured interviews, which were modified as the study progressed. Data analysis included verification of the veracity of findings by another investigator who served as an auditor. The results of this study revealed new insights into the components of client expectations that are likely to influence satisfaction with the genetic counseling process. These findings may now form the basis of hypotheses that could be tested in a quantitative study in which expectations could be measured, described or defined in measures to evaluate the effectiveness of genetic counseling services.

When the intent is to increase sensitivity, qualitative methods may reveal a rich description of an experience that is relatively novel in healthcare. This was the case in the study conducted by Williams *et al.* (2000), in which individuals who serve as support persons throughout the predictive testing experience for HD were interviewed to identify common concerns and expectations. This descriptive qualitative study described three categories of concerns for support persons. Although the results of this type of study would not be generalizable to all clinical situations, one genetics practitioner used these findings to modify a clinical protocol, allowing her to test the efficacy of an assessment of persons who serve as support persons.

An example of the third use of qualitative research is the gathering of data to develop an instrument. Skirton used results of qualitative studies for this purpose. An instrument to measure satisfaction with genetics counseling services includes items from data obtained in Skirton's (2001) grounded theory study. Instruments developed from qualitative research undergo psychometric testing before being used in quantitative research studies or for clinical purposes.

Not all research studies will fall into either a qualitative or quantitative category. In these instances, researchers use elements of both to create a mixed-method study. Generally, the study is designed to meet research criteria for one category, but a data collection method from the other category may be added. This is what McConkie-Rossell *et al.* (2001) did in their longitudinal study of perceptions and coping regarding carrier testing for Fragile-X syndrome among at-risk women. This study obtained data both from a visual analog scale that yielded a numerical rating, as well as interviews. These data were used to determine changes in responses to carrier-testing information, as well as major themes that described how participants coped.

7.5 Relevant research methods: quantitative

Quantitative methods are used when the intent is to describe, compare, correlate or evaluate relationships or interactions among variables (Burns and Grove, 2001). When the intent is to produce descriptive data, a researcher measures specific variables in a particular population. For example, in 2000, Benjamin and Lashwood conducted a study to describe characteristics and to test request patterns from individuals who had a 25% risk for HD. In this study, findings were reported as frequencies, with statistical analyses being used to identify significant differences between variables in this one population, such as participant's age or if the participant's parent was living.

In some instances, researchers will also want to compare descriptive data between naturally existing populations. These studies may identify disease-specific groups or healthcare-provider groups as the population of interest. A comparison of intentions and attitudes between those at risk for colorectal cancer who did or did not participate in genetic counseling and testing was conducted in a comparative descriptive study (Keller *et al.*, 2004). Comparisons between these naturally occurring groups suggested that intent and attitude toward genetic testing did not predict testing behaviors. This research method is also used when a researcher wants to compare outcomes of a component of healthcare provided by different practitioners. Bernhardt and colleagues (2000) used a comparative descriptive design to document no statistically significant differences in knowledge regarding breast cancer susceptibility testing when the information was provided by genetic counselors or by nurses.

The intent of research may also be to seek correlations between important variables within a single population. When this is the concern, the researcher is not seeking to identify differences between groups (descriptive category) or to identify a cause-and-effect relationship (experimental category), but rather to identify a relationship or association among variables. An example of this type of study is an epidemiologic inquiry in which it is believed that an association exists between variables within a population that is not due to chance alone. Armstrong and colleagues (2000) detected significant relationships among factors involved in the decisions to use genetic testing in a population of women who were at risk for hereditary breast or ovarian cancer. In contrast to the study conducted by Keller *et al.* (2004), Armstrong's study did not compare two groups (those who did or did not test), but identified factors, such as wanting cancer risk information for family members, that were significantly associated within a group who completed clinical genetic testing.

Researchers also conduct studies in which they wish to test hypotheses, or seek a cause-and-effect relationship. These are commonly referred to as experimental studies. In these instances, there are specific variables that can be measured. Interventions, or treatments, are administered in a consistent and controlled manner. The results are compared between participants in the experimental group and those in a comparison or control group. An example is a study in which decision-related outcomes were compared between a group of women being tested for the *BRCA1/2* mutation who received a decision aid, and a control group who did not (van Roosemalen *et al.*, 2004). Similarities or differences in decision-related outcomes were reported for the two groups. A similar type of research design was used to test the effectiveness of two education approaches, using either a genetic counselor or an interactive computer program to investigate breast cancer susceptibility (Green *et al.*, 2001). In this study, one group received both interventions, and specific outcomes from the two interventions were compared. In each of these studies, an experimental research method most closely fit the questions to be answered in the study.

7.6 Ethical considerations in research

Specific principles address the protection of human subjects when they participate in research. In the USA, three principles were identified in the Belmont Report (US Department of Health, Education, and Welfare, 1979), which identifies basic ethical principles and provides guidelines to assist in resolving ethical problems that emerge in the conduct of research with human subjects. These principles are respect for individuals, beneficence, and justice. Each of these may have many dimensions and subdivisions.

The principle of respect for individuals addresses the belief that individuals should be treated as autonomous agents and those with diminished autonomy are entitled to protection (US Department of Health, Education, and Welfare, 1979). This principle provides the foundation for policies governing the process of informed consent, freedom from coercion, and procedures designed to protect specific groups that are regarded as vulnerable populations. Examples of these groups are minor-aged children, prisoners or persons with diminished cognitive capacity.

The second principle, beneficence, refers to an obligation to do no harm, and to maximize the possible benefits while minimizing potential harm (US Department of Health, Education, and Welfare, 1979). Elements of informed consent address this principle – for example, the identification of alternative procedures that are available to individuals who are asked to enroll in a clinical trial. The third principle, justice, refers to the belief that all persons must be treated equally. In a research study, this would mean that each person would have an equal opportunity to enroll in a study and that no class or person (for example, individuals in an institution) should be singled out to become research participants.

Professional organizations have established codes of ethics that provide frameworks for the ethical conduct for practice. Examples of these are the Code of Ethics for Nurses, issued by the American Nurses Association, the Code of Ethics of the Association of Genetic Nurses and Counsellors, and the National Society of Genetic Counselors Code of Ethics (American Nurses Association, 2001; AGNC, 2004; NSGC, 1992). Genetic healthcare professionals should be familiar with the ethical principles that guide their practice, and be able to apply these to research situations. Several aspects of research require this understanding.

7.7 Consent issues

Informed consent means that individuals, to the degree that they are able, are given the opportunity to choose what shall or shall not happen to them (US Department of Health, Education, and Welfare, 1979). One of the issues in the informed consent process for research studies is that there are differences between clinical procedures and those that are part of research. For example, if the research study included an experimental treatment, or a search for genetic factors associated with a trait of interest, the participant may or may not receive results from the study, and the results may not be immediately useful to the participant. When a study includes an experimental treatment, the likelihood of benefit or harm from the treatment may not be completely known; therefore, all individuals who receive the treatment may not have an improvement in their health or well-being.

Additional issues include the types of information that can be revealed, disposition of research samples, and the range of potential personal, family or social adverse sequelae from the research (ASHG Report, 1996). The opportunity for dialog between the participant and the researcher is pivotal for accomplishing informed decision-making that is a component of informed consent (ISONG, 2000). This informed consent process has been described as the single most important mechanism for ensuring protection of individuals partaking in research from unrealistic expectations and from other potential harms of being involved in research participation (Friedmann, 2000).

In addition to the need for the opportunity for the research participant to have access to complete information, informed consent also relies on a process that supports comprehension of information. This may be an issue when members of a vulnerable population qualify for enrollment in a research study. Although specific definitions of what constitutes a vulnerable population may differ, examples of groups that may require accommodations to promote comprehension are persons with conditions limiting communication, individuals who are reading a consent that has been translated from one language to another, those with cognitive impairment or psychiatric disturbances, people at the end of life or persons from minority populations (ISONG, 2002). Allowing adequate time, opportunity for information to be provided in a way that is understandable to the person considering participating in research and freedom from perceptions of coercion, all contribute to helping people make an informed decision. The Genetic Alliance (2001) has published an excellent brochure for members of the public to assist them in understanding components of the informed consent process for genetic research studies.

Minor-aged children

One other category of potential research subjects may also require special consideration – the category of minor-aged children. There are many resources that address ethical and clinical questions regarding genetic testing of children for clinical purposes. In addition to these questions, other issues must be considered when minor children may be asked to participate in research. By virtue of their developmental stage and lack of autonomy, children are regarded as a vulnerable population with regard to participation in research (Broome, 1999). In the USA, the National Institutes of Health (NIH) stated that research investigators should include children in their research or document why that would not be appropriate (NIH, 1998). However, children may not be able to completely understand the purposes and potential risks or benefits of research, and dependent minor-age children cannot provide consent for participation in research.

Guidelines for ethical conduct of medical research involving children emphasize that the children's interests must be assessed. In addition, established ethics guidelines for any research study must be followed (McIntosh et al., 2000). Several considerations should be weighed when considering the enrollment of children. In addition to adapting information to the child's cognitive level, efforts should be made to support the child's ability to give his/her own opinion about participation in research. A term that is used in clinical research is 'assent.' This refers to the permission or agreement given by a child, who is 7 years of age or older (Broome, 1999). Early research noted that, for some children, the notion of making a free choice might be overshadowed by preferences or decisions made by parents or healthcare providers (Scherer, 1991; Susman, 1992). However, a more recent study (James *et al.*, 2003) – on the perceptions of children and their parents about carrier testing – provides insights into children's feelings about genetic testing and their feelings about discussing their questions with their parents.

7.8 Dissemination of information, presentations, posters, writing for publication

Many research studies are conducted by teams, of which genetic healthcare practitioners are members. The final phase of a research study is the reporting of the results. This may include publication in printed or internet journals, as well as via oral presentations or posters at professional meetings. One of the relevant decisions concerns who becomes an author on research reports. In some cases, the journal to which the article is being submitted has guidelines, and the editor asks authors to identify to which components they contributed. Guidelines to determine authorship are issued by some professional organizations (Nativio, 2000). In still other instances, it is a decision that is made by the lead researcher. Regardless of the process, it is helpful if members of the research team can discuss rules or expectations about authorship early in the research process.

Other considerations in preparing a manuscript involve the process of writing. One excellent guide is a paper by Bowen (2003), who provides clear step-by-step directions for assembling a research report for publication. One aspect of preparing reports of genetic research is protection of privacy of any participants in the study. Topics of special concern may include family pedigrees, descriptions of clinical data, or other descriptive information that could identify an individual or a family. Some journals that publish genetically focused manuscripts are developing guidelines that specifically address privacy protection. In addition to concerns about privacy, journal editors may also ask authors to identify any conflicts of interest, such as serving on a board or as a consultant for a manufacturer or drug company whose product was tested in the research study.

Results of research are also presented via posters or platform presentations at professional meetings. The time between completion of the research and the meeting may be shorter than the time between conclusion of a study and the publication of results in peer-reviewed journals. Just as with published findings, research presentations or posters should be well-organized, providing proper acknowledgment of all contributors and funding sources. Helpful hints for preparing slides for platform presentations include using no more than eight lines per slide and no smaller than 32-point font, avoiding red/green contrast and using only horizontal layout. One of the best ways to prepare any report for dissemination via publication, presentation or poster is to ask peers to review and critique the report to identify components that are unclear, redundant, or incompletely presented.

Genetic healthcare professionals may also be asked to speak to the media about the results of a research project. This is an important opportunity to educate the public regarding the findings of the research and how these findings contribute to knowledge about a healthcare problem. Preparing for a media interview includes being available when a member of the media requests time for the report, being prepared to speak succinctly and in plain language, and knowing what information you want to provide to the reporter and the public (Farberman, 2000).

7.9 Conclusion

Research is the process by which new knowledge is generated and existing knowledge is expanded. Genetics practitioners participate in research through the identification of important research questions, through the design and implementation of research studies and the reporting of the results. Ethical principles that protect the rights of human subjects who participate in research provide a foundation for the ethical conduct of research on genetic topics. Dissemination of research allows peers, other professionals and the public to critique and benefit from the findings of research.

Study Questions

Case scenario – Ben

Ben, who is 13 years old, is the son of Judy and Roger. Ben is one of five children; his sisters are Amy (10 years old) and Kathleen (16 years old). Ben had two older brothers – Roy (who died when he was 15 years old) and John (who died when he was 17 years old) – but both died about 3 years ago. Ben, Roger and John all have a diagnosis of Duchenne muscular dystrophy (DMD). Judy has two sisters, Barb and Nancy. Barb has two daughters who are healthy. Nancy has two daughters and two sons, all of whom are alive and well. Judy's mother, Martha, had two brothers – William (who died when he was 12 years old) and Gordon (who died when he was 14 years old), both of whom died of muscle weakness; their medical history is consistent with DMD.

Task A

Draw a basic family tree.

Task B

Ben receives medical care from a specialty clinic that includes a neurologist, genetics health professional, physical therapist and social worker on the clinic team. He was diagnosed 6 years ago. His family does not consistently keep their appointments, and the last time Ben was seen was a year and a half ago. As a part of the clinic program, there is a new research study being developed to examine the impact of two counseling approaches for discussion of end-of-life decisions. The genetics health professional is a member of the research team. The inclusion criteria for this study are parents and their teenage children with DMD. One of your roles, as a member of the research team, is to communicate with individuals who qualify for the study and enroll them in the study. You contact Judy to discuss the research project with her.

(a) Judy is not certain if she wants to participate in the study, but she tells you she doesn't want to upset Ben's doctor, so she thinks she will participate in the study. She also thinks Ben should participate, and she tells you she will sign his consent for him. What ethical principles may guide your response to Judy's statement?

(b) The study includes two treatment groups for the parents, one in which parents receive a brochure and individual discussion with a member of the clinic team. The other group receives a brochure and participation in a parent support group. Judy tells you she prefers to be in the brochure and individual discussion group, as she does not like to talk about these topics when other people are around. What would be the consequences if Judy could select which treatment she received? What ethical principles provide guidance for the research plan to randomize participants to either group? What information would you include in your response?

(c) Following completion of the study, you contact the research team regarding reporting of the results. You have been asked by your professional organization to present the results at their annual meeting that will be held in 4 months time. The director of your clinical unit tells you that he/she should be listed as the first author on all reports from this study. What resources might help you in making authorship decisions?

References

AGNC (2004) Code of Ethics. [Accessed March 3, 2005] http://www.agnc.co.uk/Registration/registration.htm

Genetic Alliance (2001) Informed consent: participation in genetic research studies. [Accessed March 3, 2005] http://www.geneticalliance.org

American Nurses Association (2001) Code of ethics for nurses with interpretive statements. [Accessed July 22, 2004] http://nursingworld.org/ethics/code/ethicscode.150.htm

Armstrong, K., Calzone, K., Stopfer, J., Fitzgerald, G., Coyne, J. and Weber, B. (2000) Factors associated with decisions about clinical BRCA1/2 testing. Cancer Epidemiol. *Biomarkers Prevent.* **9**: 1251–1254.

ASHG Report (1996) Statement on informed consent for genetic research. *Am. J. Hum. Genet.* **59**: 471–474.

Barnes, C. (1998) Testing children for balanced chromosomal translocations: parental views and experiences. In: *The Genetic Testing of Children*, (ed. A. Clarke) BIOS Scientific Publishers, Oxford, pp. 51–60.

Benjamin, C.M. and Lashwood, A. (2000) United Kingdom's experience with presymptomatic testing of individuals at 25% risk for Huntington's disease. *Clin. Genet.* **58**: 41–49.

Bernhardt, B.A., Geller, G., Doksum, T. and Metz, S.A. (2000) Evaluation of nurses and genetic counselors as providers of education about breast cancer susceptibility testing. *Oncol. Nurs. Forum* **27**: 33–39.

Bernhardt, B.A., Tambor, E.S., Fraser, G., Wissow, L.S. and Geller, G. (2003) Parents' and children's attitudes toward the enrollment of minors in genetic susceptibility research: implications for informed consent. *Am. J. Med. Genet.* **116A**: 315–323.

Bowen, N.K. (2003) How to write a research article for the *Journal of Genetic Counseling*. *J. Genet. Counsel.* **12**: 5–21.

Broome, M.E. (1999) Consent (assent) for research with pediatric patients. *Semin. Oncol. Nurs.* **15**: 96–103.

Burns, N. and Grove, S. (2001) The Practice of Nursing Research. 4th Edn. W.B. Saunders, Philadelphia, PA.

Farberman, R.K. (2000) Preparing for media interviews. *Monitor on Psychol.* **31**(10): 68–69.

Friedmann, T. (2000) Principles for human gene therapy studies. *Science* **287**: 2163–2164.

Gordon, C., Walpole, I. Zubrick, S.R. and Bower, C. (2003) Population screening for cystic fibrosis: knowledge and emotional consequences 18 months later. *Am. J. Med. Genet.* **120A**: 199–208.

Green, M.J., McInerney, A.M., Biesecker, B.B. and Fost, N. (2001) Education about genetic testing for breast cancer susceptibility: patient preferences for a computer program or a genetic counselor. *Am. J. Med. Genet.* **103**: 24–31.

ISONG (2000) Position statement: informed decision-making and consent: the role of nursing. [Accessed July 23, 2004] http://www.isong.org

ISONG (2002) Position statement: genetic counseling for vulnerable populations: the role of nursing. [Accessed July 16, 2004] http://www.isong.org

ISONG (2004) What is a genetics nurse? [Accessed December 12, 2004] http://www.isong.org

James, C.A., Holtzman, N.A. and Hadley, D.W. (2003) Perceptions of reproductive risk and carrier testing among adolescent sisters of males with chronic granulomatous disease. *Am. J. Med. Genet.* **119C**: 60–69.

Keller, M., Jost, R., Kadmaon, M. *et al.* (2004) Acceptance of and attitude toward genetic testing for hereditary nonpolyposis colorectal cancer: a comparison of participants and nonparticipants in genetic counseling. *Dis. Colon Rectum* **47**: 153–162.

Knafl, K.A. and Webster, D.C. (1988) Managing and analyzing qualitative data: a description of tasks, techniques, and materials. *Western J. Nurs. Res.* **10**: 195–218.

McConkie-Rosell, A., Spiridigliozzi, G.A., Sullivan, J.A., Dawons, D.V. and Lachiewicz, A.M. (2001) Longitudinal study of the carrier testing process for Fragile-X syndrome: perceptions and coping. *Am. J. Med. Genet.* **98**: 37–45.

McIntosh, N., Bates, P., Brykczynska, G. *et al.* (2000) Guidelines for the ethical conduct of medical research involving children. *Arch. Dis. Children* **82**: 177–182.

Nativio, D. (2000) Authorship. *J. Nurs. Scholarship* **12**: 351.

NIH (1998). NIH policy and guidelines on the inclusion of children as participants in research involving human subjects. [Accessed March, 6 1998]. Bethesda MD. NIH [online] http://grants.nih.gov/grants/guide/notice-files/not98-024.html

NSGC Code of Ethics (1992) [Accessed July 22, 2004] http://www.nsgc.org/newsroom/codeofethics.asp

Scherer, D. (1991) The capacities of minors to exercise voluntaries in medical treatment decisions. *Law Hum. Behav.* **15**: 431–449.

Skirton, H., Barnes, C., Guilbert, P., Kershaw, A., Kerzin-Storrar, L., Patch, C., Curtis, G. and Walford-Moore, J. (1998) Recommendations for education and training of genetic nurses and counsellors in the United Kingdom. *J. Med. Genet.* **35**: 410–412.

Skirton, H. (2001) The client's perspective of genetic counseling: a grounded theory study. *J. Genet. Counsel.* **10**: 311–329.

Susman, E., Dorn, L. and Fletcher, J. (1992) Participation in biomedical research: the consent process as viewed by children, adolescents, young adults, and physicians. *J. Pediatr.* **124**: 547–552.

US Department of Health, Education, and Welfare. (1979) Protection of human subjects: Belmont Report – Ethical principles and guidelines for the protection of human subjects of research. *Fed. Regist.* **44**: 23192–23197.

Van Roosmalen, M.S., Stalmeier, P.F., Verhoef, L.C., Hoekstra-Weegers, J.E., Oostewijk, J.C., Hoogerbrugge, N., Moog, U. and van Daal, W.A. (2004)

Randomised trial of a decision aid and its timing for women being tested for a BRCA1/2 mutation. *Br. J. Cancer* **90**: 333–342.

Williams, J.K., Schutte, D.L., Holkup, P.A., Evers, C. and Muilenburg, A. (2000) Psychosocial impact of predictive testing for Huntington disease on support persons. *Am. J. Med. Genet.* **96**: 353–359.

Further resources

Brink, P.J. and Wood, M.J. (1998) *Advanced Design in Nursing Research.* 2nd Edn. Sage, Thousand Oaks, CA.

Bruning, J.L and Kintz, B.L. (1987) *Computational Handbook of Statistics.* 3rd Edn. Scott, Foresman and Company, Glenview, IL.

Burns, N. and Grove, S. (2001) *The Practice of Nursing Research.* 4th Edn. W.B. Saunders, Philadelphia, PA.

Coffey, A. and Atkinson, P. (1996) *Making Sense of Qualitative Data.* Sage, Thousand Oaks, CA.

Girden, E.R. (1996) *Evaluating Research Articles from Start to Finish.* Sage, Thousand Oaks, CA.

Uniform Regulations for manuscripts submitted to biomedical journals. [Accessed August 2, 2004] http://ww.icmje.org

8 Autosomal dominant inheritance

8.1 Definition of dominant inheritance

The inheritance of traits from parent to child has been of interest to humans for thousands of years, and the search for features that can link a newborn infant with those of the family's ancestors is a common experience in many Western families. In 1865, the Austrian monk Gregor Mendel identified several principles of inheritance, including the principle of Independent Assortment, which states that genes at difference loci are transmitted independently. A second principle, the Principle of Segregation, also developed by Mendel, states that sexually reproducing organisms possess genes that occur in pairs and only one member of the pair is transmitted to each offspring (Jorde et al., 2003), Mendel's work led to the identification that some alleles behave in a dominant manner, whereas others behave in a recessive manner (see 'Further resources' for a more complete review of this subject). In 1909, the term 'gene' was coined by Johannsen to describe the basic unit of heredity (Jorde et al., 2003). These principles underlie the understanding of autosomal dominant inheritance, which is one pattern of inheritance that is dependent on the basic unit of heredity. Although current understanding is continuing to be revised of both the inheritance of genes and also the potentially complex explanations for the influence of genetic and environmental factors on health and disease, much of our understanding of these influences are built on the foundations of Mendel's laws of inheritance (Powledge, 2001).

8.2 Dominant inheritance pattern

Dominant inheritance occurs when one member of a gene pair contains a mutation in that gene, and this mutation is associated with symptoms of a condition or a disease. The genes that are associated with most dominantly inherited traits are located on the autosomes; they are therefore referred to as autosomal dominant traits. Of the single gene conditions (more than 6000 of which have been identified), more than half are autosomal-dominant traits (Nussbaum *et al.*, 2001). The term 'phenotype' describes the characteristics that can be observed or measured, and it cannot be assumed that there is a direct relationship between a person's genotype and their phenotype.

Features of autosomal dominant inheritance

Autosomal dominant inheritance involves the following features:

- vertical transmission through males and females;
- unless it is a new mutation, an affected person will have an affected parent;
- either sex may be affected;
- an affected parent has a 1 in 2 or 50% chance of having an affected child.

CASE STUDY – REBECCA

Rebecca Talbot is a 24-year-old woman who is married and who has two children. She is the accounts manager for a manufacturing plant. Rebecca has two sisters – Tracy, age 22, and Joanna, age 19. Her mother, Trish Martin, was always regarded as having a strong personality, and she did not get along with her own parents and siblings. Trish left home when she finished high school, and maintained limited contact with her aunt Jane, age 67, one of Trish's mother's three sisters.

Trish, when she was 45 years old, died in an automobile accident. While at her mother's funeral service, Rebecca met her great-aunt Jane for the first time. During their conversation, Jane said, 'It was a blessing she didn't get what the rest of them suffered with'. Rebecca learned that her mother's father, Pete, and two of her mother's brothers, Richard and Steven, had been diagnosed with Huntington disease (HD). Pete died in a hunting accident when he was in his late 30s. Both of Trish's brothers have also died: one had taken his own life and the other died after living in a nursing home for 12 years.

Rebecca was shocked by this family news. She had not been aware of this information before her mother's death and was already having a difficult time dealing with the sudden loss of her mother. She conducted a search on the internet for information on HD. Here, she found information about the disease, research studies, personal stories regarding genetic discrimination and the availability of predictive testing. Rebecca did not share this information with anyone. After completing her internet search, she contacted the medical genetics unit and requested an appointment for predictive testing for HD. When she talked with the genetics nurse, she gave her name as Trish Martin.

8.3 The distinction between risk of inheritance and risk of disease

Communication of risk information in the context of dominantly inherited conditions can involve two types of information: the likelihood that a gene mutation was transmitted to an individual from the parent, and the chance that the person who inherits the gene mutation will develop the condition or disease. Risk information about inheritance of an autosomal dominant gene mutation may be described as a fraction (½ chance), an odds ratio (1:1) or as a percentage statement (50% chance). This likelihood may be perceived as being higher when the information is presented as a fraction (½) rather than as a percentage risk (50%) (Kessler and Levine, 1987; Michie and Marteau, 1996). Risk information about the likelihood that a person will develop the disease is a more disease specific calculation and depends on several pieces of information, including:

- whether a specific gene mutation has been identified for that condition in that person;
- what is known regarding the penetrance of the gene mutation;
- other mitigating factors that may influence emergence of symptoms.

8.4 Penetrance

A component of communication of risk information refers to the likelihood that a person with a gene mutation will develop the phenotype. It is recognized that genes interact with each other and with the environment, which requires caution when applying Mendel's original laws to diseases with a genetic component. One feature of inherited traits is penetrance. This term has traditionally been based on population data and refers to the proportion of obligate carriers of dominantly inherited disorders who demonstrate signs of the disease. With discoveries from research into diseases that may result from a combination of genetic and environmental factors, this term may also become more commonly used to refer to contributions of genes for common chronic diseases. When there is reduced, or incomplete penetrance, an individual has the mutation, but does not manifest signs of the disease. This may be determined from the person's position in the family pedigree, or from molecular genetic testing. For example, a woman may be a member of a family in which she (third generation) and her maternal grandmother (first generation) have both been found to have a mutation in the BRCA1 gene for hereditary breast and ovarian cancer (HBOC). However, her mother, (in the second generation) is recognized to be an obligate carrier, but she does not have breast or ovarian cancer.

8.5 Expressivity

Another important characteristic of some inherited traits is variability in the severity of the condition; this is known as expressivity or variable expression. In some disorders, there may be a high degree of penetrance (that is, most people with the mutation demonstrate some signs of the condition), but individuals may have features of the condition in a more severe or a milder form. This has been observed in individuals who have neurofibromatosis type I (or von Recklinghausen) disease. People with this disorder may have only a few café-au-lait spots, whereas others have more severe expression of the disorder, including multiple neurofibromata, malignant sarcomas or learning disabilities (Jorde et al., 2003; Nussbaum et al., 2001). For some conditions, specific mutations may be associated with various clinical features of a disease. However, genotype/phenotype associations have not yet been identified in the gene NF1. Therefore, results of molecular DNA testing would not be helpful in predicting which clinical symptoms a person might have, or how serious they could be.

8.6 Allelic heterogeneity

In other disorders, some clinical information can be derived from identification of the specific mutation present in an individual. In these examples, variable expression may be associated with different types of mutation at the same disease locus. This is termed allelic heterogeneity. For example, some mutations in the gene that causes adult polycystic kidney disease may also cause berry aneurysms (Harper, 2004). In addition to studies of gene and protein function, environmental factors may eventually be identified that will provide insights into these observations. It is also possible that a collection of symptoms may result from mutations in genes on several chromosomes, with various patterns of inheritance. This is termed locus heterogeneity. For example, loci on more than 20 chromosomal regions have been identified in retinitis pigmentosa (RP), a group of diseases that are associated with retinal degeneration, one of the most common inherited causes of blindness (Jorde et al., 2003). Twenty-five percent of cases of RP are inherited in an autosomal dominant pattern, with another 30% being acquired via X-linked inheritance and 45% via autosomal recessive inheritance (Nussbaum et al., 2001).

CASE STUDY – REBECCA

In Rebecca's family, the diagnosis of Huntington disease (HD) was confirmed in several family members. Her family pedigree illustrates autosomal dominant inheritance through the transmission of the mutation for HD from parent to child (Pete to Richard, Steven, and possibly to Trish). It also illustrates the presence of the condition in both males and females, as well as the disease phenotype in two generations.

Rebecca is desperate to find out if she inherited the gene mutation for HD, and if so, if she will definitely develop the disease. She also wants to know if she could have already passed it on to her own daughters.

8.7 Types of mutation

The types of potential gene mutation have been described in Chapter 2. When mutations alter individual genes, they may be classified by their structural description or functional consequence (Guttmacher *et al.*, 2004). Types of point mutation include missense, nonsense, and frameshift mutations. A missense mutation occurs when a different base pair results in the creation of a different amino acid, rather than in the one that was originally specified. Nonsense mutations, which account for approximately 12% of all disease-causing mutations (Nussbaum *et al.*, 2001) occur when the alteration results in the cell stopping the production of a new protein. A frameshift mutation occurs when an additional base or a removal of a base or bases changes the reading frame.

8.8 Functional categories of mutations

Mutations can also be classified by their functional effect. Some have a neutral effect. This category includes the silent mutations in which one base is replaced with another that codes for the same amino acid (Guttmacher *et al.*, 2004). Some may be associated with disease, whereas others have a beneficial effect. In the case of sickle-cell disease (SCD), which is an autosomal-recessive disorder, a single missense mutation causes a substitution of the amino acid valine for the amino acid glutamic acid. A person who is homozygous for the mutation at this locus has the clinical disorder, SCD. However, in the heterozygous state, heterozygote advantage can occur, in which carriers demonstrate resistance to malaria (Nussbaum *et al.*, 2001).

In some diseases, more than one type of mutation has been reported in individuals who have the disease phenotype. An example is the finding of substitutions, deletions, insertions, and splice-site mutations in a group of individuals with NF1 (Origone *et al.*, 2003).

Of the mutations that lead to disease, the most common consequence is a 'loss of function.' In this case, the protein that is normally made by the cell is made in a smaller amount, or the product does not function properly. An example of a loss of function is the abnormal enzyme that is produced in the disease glucose-6-phosphate dehydrogenase (G6PD) deficiency, an X-linked recessive disorder, in which the male with this gene mutation can develop an acute hemolytic anemia when exposed to certain drugs (Guttmacher *et al.*, 2004). In other instances, a gene mutation can result in a gain of function, or too much of a protein. One inherited form of peripheral nervous system disease is Charcot Marie Tooth disease, in which autosomal dominant, autosomal-recessive and X-linked recessive inheritance have been reported. In this condition, a gain of function occurs when there is a duplication of a portion of the gene, leading to an increase in the amount of gene product made, and thereafter resulting in the clinical symptoms of the disorder (Jorde *et al.*, 2003).

In HD, the expanded repeat is located within the coding region of the gene. The abnormally large number of repeats leads to the production of an altered version of the huntingtin protein (Trottier and Mandel, 2001).

Expansion

When the Mendelian patterns of inheritance were identified, it was believed that genetic material was passed, unchanged, from parent to offspring. This assumption was revised when it was discovered that the number of trinucleotide repeats (such as in the gene for HD) was found to increase when passed from an affected parent to his or her offspring. Known as 'anticipation', this phenomenon can lead to progressively earlier onset and increased severity of these diseases within successive generations in a family (Nussbaum *et al.* 2001). When a mutation has no phenotypic effect, but has the capacity to increase in size during meiosis, this is termed a premutation. The extent of expansion that leads to disease is specific for each condition in which expansion occurs, and may be associated with the gender of the parent in whom the premutation is present. For example, in HD, the normal range of trinucleotide repeats is 19–35, with more than 36 repeats being observed in people with the disease (Menalled and Chesselet, 2002). The expansion of the CAG trinucleotide repeat in the HD gene mutation is observed most frequently when this mutation is passed from a father to his offspring. In other conditions, such as the fragile X syndrome, individuals with this disorder have 200 or more trinucleotide repeats in a gene on the X chromosome. Expansion of this gene mutation is observed when the gene mutation is passed from a mother to her offspring.

8.9 Predictive testing

Purposes

When the location of the HD gene was identified in 1993, guidelines were developed for the administration of predictive testing for individuals at risk of developing HD (HDSA, 1994; Went *et al.*, 1994). These guidelines have been used as a model for predictive testing for other inherited conditions, particularly neurodegenerative disorders or familial cancer syndromes (see 'Further resources'). Elements of the guidelines address use of genetic testing, confidentiality, support services for the person requesting testing and management of special situations. Issues regarding analytic and clinical validity, as well as quality assurance regarding the laboratory process, are also considered by genetics services providers when a test is requested for predictive purposes. Genetests (http://genetests.org) is an excellent resource from which genetic healthcare providers can obtain accurate and current information about tests and laboratories that provide them.

One of the purposes of genetic testing is to predict the likelihood that an individual who is at risk for an inherited condition will develop the disease. The protocol for predictive testing for neurodegenerative disorders (such as HD, cerebral autosomal dominant arteriopathy with subcortical infarcts and leukoencephalopathy (CADASIL) or dentatorubral-pallidoluysian atrophy (DRPLA)) specifies that adequate time be provided to complete genetic counseling, neurologic and psychological assessments, and discussion of considerations within the informed consent process. The protocol also specifies that the results be reported in person, and that follow-up contact be made available.

Limitations

One of the limitations of predictive testing can be the accuracy of the test. Analytical validity is a term used to describe how well a test measures the property it intends to measure. For conditions in which there is one known mutation that is easily detectable – for example, an expanded fragment – the analytical validity will be high. However, in some conditions there are many mutations in one gene, or mutations in many genes, some of which are not yet known.

Discussions of limitations of the test also include addressing if the test results are associated with treatment, surveillance or prevention strategies. This is often the case when testing for cancer predisposition, in which surveillance and/or prophylactic surgery may be feasible. However, in the neurodegenerative conditions, the test results are not yet useful in making treatment decisions about management of symptoms. For example, in HD, the number of repeats is associated with age of onset, with earlier age of onset being associated with larger repeat sizes (Djousse, L. *et al.*, 2003). However, the results of the test do not provide specific information regarding the timing or nature of symptoms that will first be experienced in a person with HD.

Predictive testing – benefits

Informed consent also includes a discussion of potential benefits and risks. Reasons why people have reported that testing was beneficial to them include the ability to inform their children if they are at risk, being relieved from the uncertainty of not knowing if they have the gene mutation, and having the option to use this information for personal decisions such as reproduction, occupation or fulfilling goals during a time of relatively good health (Codori and Brandt, 1994; van't Spijker and ten Kroode, 1997; Wahlin *et al.*, 1997). Those who learn they do not have the gene mutation for HD report that they can stop monitoring themselves for the initial signs of the disease, and that they are relieved that their children are not at risk of developing this disorder (Williams *et al.*, 2000).

Predictive testing – risks for harm

Learning that one does or does not have the mutation for HD cannot be undone once genetic testing is completed and the results are reported to the person having the test. Some individuals who have had a predictive test note that their lives are now changed and that they cannot return to the assumptions they had made about themselves before the testing (Williams and Sobel, in press). For some this is distressing, and can lead to alterations in their own emotional health and relationships with others. Risks for emotional distress also include possibilities for depression, inability to avoid thinking about HD or more serious consequences. One long-term study reported that as people who had a positive predictive HD test neared the time of onset of HD, their emotional distress increased (Timman *et al.*, 2004). The likelihood of adverse psychological events has also been monitored in individuals who complete predictive testing for HD. Defined as suicide, suicide attempt, and psychiatric hospitalization, adverse events were documented in less than 1% of persons in a worldwide cohort who had a positive HD test (Almqvist *et al.*, 1999). Risk factors that were identified in this study included a psychiatric history of less than 5 years before testing and being unemployed.

Issues about predictive testing also include the potential impact that the results will have on the person's personal and family relationships. Individuals who have completed predictive testing and have been found not to have the gene mutation report changes in their relationships with some family members, especially those who have not been tested or who have signs of HD. Those who learn that they will develop HD may experience loss of being regarded as a leader in the family, and those who learn that they will not develop HD may feel estranged and no longer share the sense of identity held by others who do not know their actual HD risk (Sobel and Cowen, 2003; Williams *et al.*, 2000).

It is clear that predictive testing is not preferred by everyone who is at risk of developing HD, with current estimates ranging from around 5% in the USA to 15–35% for individuals from Europe, Canada or Australia (Creighton *et al.*, 2003). One area that is identified as a potential risk, is the risk for genetic discrimination. In the USA, individuals who complete genetic testing note that risk of loss of insurance is one deterrent in considering whether or not to have predictive testing. People worry that they may experience discrimination in the workplace or in obtaining health insurance (Penziner, 2004). Although in most states that prohibit insurance companies from using genetic information to deny coverage or to raise health insurance rates, laws have been passed. However, once a person develops symptoms, these laws do not apply (Rothstein, 1999). Individuals who request predictive genetic testing may prefer to not have their name recorded on laboratory or medical record documents, and conceal their names in an attempt to avoid risk for potential insurance or other genetic discrimination.

In the UK, where there is a publicly funded healthcare system, concerns about discrimination relate primarily to life insurance. Most people require life insurance as a condition of their mortgage. Discussions between the insurance industry, the genetics community and the Department of Health led to a moratorium on the requirement for genetic test results to be disclosed to insurance companies. This is subject to a financial limit and is due to be reviewed in 2006.

CASE STUDY – REBECCA

When Rebecca Talbot contacted the genetics unit, she requested predictive testing for Huntington disease (HD). As part of the testing process, the genetic healthcare practitioner discussed aspects of the test, including the purposes, limitations, benefits and potential harms of the test with Rebecca and her support person. The practitioner also helped Rebecca to focus on the meaning of a positive or negative result for herself, and for her children.

Rebecca informed the counselor that she was mainly interested in the test because she felt she needed to inform her children of their risk as soon as possible. Deep discussion between Rebecca and her practitioner revealed that Rebecca was very angry about the fact that she had not been told of her risk by her parents. She was afraid that her own children would resent her if she did not inform them at a young age.

One of the recommendations for predictive testing is that more than one appointment is offered with the genetics team, so that information can be introduced, reviewed and discussed prior to testing. It is also recommended that a support person accompany a person seeking predictive testing for HD. In Rebecca's case, her husband came with her to her appointments with the genetics counseling team. As the family history was recorded, it became apparent that Rebecca was not using her own name and the reasons for this were explored with Rebecca, who said that she was afraid of the repercussions, should the test be positive. She was particularly concerned that her job might be at risk and that she would no longer be able to obtain health insurance. The practitioner suggested that Rebecca think more about her reasons for testing and the way she would cope with the result before returning for another appointment in a month, especially in view of her recent bereavement. She also pointed out that as Rebecca's children are both under 4 years old, any decision about telling them can be delayed for at least several years. A month later, Rebecca and her husband told the team that they have decided to delay testing until Rebecca feels she has recovered from her mother's death. She asked if she could keep in touch with the clinic, and that she would call when the time was right.

8.10 Conclusion

Autosomal dominant inheritance is present when one member of a gene pair has a mutation. Phenotypic effects of a gene mutation reflect interactions between genetic and environmental factors. For some autosomal-dominant conditions, gene mutations have been identified and testing may contribute to clinical decision-making. Genetic testing that is conducted for predictive purposes requires careful evaluation by the person at risk of developing the disease. This allows the person to weigh up the potential benefits and risks, as well as to understand the limitations of testing.

Study questions

You have received a referral letter. It reads:

Dear Genetics Team Member,

Re: Bob Woods

Mr Woods, age 23, was seen by me today, for his first visit in the Community Family Practice clinic. Three weeks ago, Bob learned that his father was diagnosed with Huntington disease (HD). He was aware of this condition in his family, as his father's sisters have this disease. He had been under the impression that only women developed this disease in their family. Mr Woods wishes to have genetic testing to rule out HD. His neurologic exam was normal today, and he does not demonstrate any signs of the typical choreic or neurologic abnormalities of this disease.

Mr. Woods is married and is the father of a 3-year-old son; his wife is 6 months pregnant.

Thank you for your attention to this request. You can contact Mr Woods at the address below.

Sincerely,
Family Nurse Practitioner

Task A
You will be calling Mr Woods to discuss this referral.

(a) Identify the goals for this initial telephone call.
(b) What questions will you want to ask him when you call?
(c) What information would you want to provide to him?

Task B

Information obtained during your telephone conversation with Mr. Woods:

He is the oldest of three siblings. His brother, Ben, is 20 years old and is a college student. His sister, Kelly, is 16 years old and in high school. Their father, Mark, is 41 years old and was diagnosed as having HD by a neurologist. Mark has three sisters and one brother. Karen is 52 years old. She has HD and has lived in a nursing home for 4 years. She has one daughter, Susan. Mark's next sister is Roseanne. She is 48 years old and is single. She was diagnosed with HD after an automobile accident 2 years ago. She lives in an assisted living facility. The third sister, Helen, is 39 years old and has two sons. Bob believes she is healthy. Bob believes Mark's brother Jerry has schizophrenia. Jerry does not maintain contact with the family and Bob does not know very much about him. Bob thinks Jerry has a son. Bob's paternal grandfather, James, is a military veteran. He was admitted to a veteran's care facility when Mark was a child. Bob remembers being told that his grandfather had a head injury during his military service, and after the children were born, he could not take care of himself. Mark's mother, Ellen, died when she was 73 years old from a stroke. Bob's mother, Gwen, is 40 years old and has diabetes. Bob has a 3-year-old son, Ryan, and his wife, Jennifer, is 6 months pregnant.

(a) Construct Mr Wood's family pedigree.
(b) Identify which family members may have HD and which are at risk of developing HD.
(c) Describe the features in this pedigree that are consistent with autosomal dominant inheritance.
(d) If these individuals are found to have HD, which family members are at a 50% risk?
(e) Which are at a 25% risk?

Task C

Mr Wood attends his first genetic testing visit. He brings Ben, who is undecided about being tested. Bob asks that they go through the testing together and that they hear everything that the other is told. Jennifer states that she will be the support person for both Bob and Ben. What issues do you consider in determining the desirability of this situation?

Task D

This is your first predictive testing client. What resources would you use to locate information on predictive testing laboratories and the meaning of potential test results?

Task E

Mr Woods is worried that his employer might find out about his having the test. What measures would you take to protect the privacy of his medical information?

Task F

You realize that Ben prefers not to have the test, but he does not want to disappoint his brother. What information would you give Bob? And what information would you give Ben?

Task G

Bob and Ben are both tested. Bob has 44 CAG repeats. Ben has 31.

(a) What are the implications of Bob's results?
(b) And what are implications of Ben's?
(c) What counseling would you offer to each of them?

References

Almqvist, E.W., Bloch, M., Brinkman, R., Crauford, D. and Hayden, M. on behalf of an international Huntington disease collaborative group (1999) A worldwide assessment of the frequency of suicide, suicide attempts, or psychiatric hospitalization after predictive testing for Huntington disease. *Am. J Hum Genet.* **64**: 1293–1304.

Codori, A. and Brandt, J. (1994). Psychological costs and benefits of predictive testing for Huntington's disease. *Am. J. Med. Genet.* **54**: 174–184.

Creighton, S., Almqvist, E.W., MacGregor, D. *et al.* (2003) Predictive, pre-natal and diagnostic genetic testing for Huntington's disease: the experience in Canada from 1987 to 2000. *Clin. Genet.* **63**: 462–675.

Djousese, L. *et al.* (2003) Interaction of normal and expanded CAG repeat sizes influences age at onset of Huntingon disease. *Am. J. Med. Genet.* **119A**: 279–282.

Guttmacher, A.E., Collins, F.S. and Drazen, J.M. (2004) *Genomic Medicine: Articles from the New England Journal of Medicine.* Johns Hopkins University Press, Baltimore, OH.

Harper, P.S. (2004) *Practical Genetic Counselling.* 6th Edn. Oxford University Press, Oxford.

HDSA (1994) [Accessed March 3, 2005] http://www.hdfoundation.org/testread/hdsatest.htm

Jorde, L.B., Carey, J.C., Bamshad, M.J. and White, R.L. (2003) *Medical Genetics,* Third Edition, St Louis, Mosby.

Kessler, S. and Levine, E.K. (1987). Psychological aspects of genetic counseling. IV. The subjective assessment of probability. *Am. J. Med. Genet.* **28**: 361–370.

Menalled, L. and Chesselet, M. (2002) Mouse models of Huntington's disease. *Trends Pharmacol. Sci.* **23**: 32–39.

Michie, S. and Marteau, T.M. (1996) Genetic counseling: some issues of theory and practice. In: *The Troubled Helix: Social and Psychological Implications of the New Human Genetics* (eds T.M. Marteau, and M. Richards). Cambridge University Press, Cambridge.

Nussbaum, R.L., McInnes, R.R. and Willard, H.F. (2001) *Thompson and Thompson Genetics in Medicine*. 6th Edn. W.B. Saunders, Philadelphia, PA.

Origone, P., Bellini, C., Sambarino, D. *et al.* (2003) Neurofibromatosis type 1 (NF1): Identification of eight unreported mutations in the NF1 gene in Italian patients. *Hum. Mutat.* **22**: 179–180.

Penziner, E. (2004) Public health implications of genetic research. Unpublished manuscript. College of Public Health, The University of Iowa, Iowa City, IW.

Powledge, T.M. (2001) *Genetic Basics*. National Institutes of Health, National Institute of General Medical Sciences, NIH Publication No. 01-662.

Rothstein, M. (1999) Protecting genetic privacy: why is it so hard to do? *Hum. Genome News* **19**: 14–15.

Sobel, S. and Cowen, D.B. (2003) Ambiguous loss and disenfranchised grief: the impact of predictive testing on the family as a system. *Family Process* **42**: 31–46.

Timman, R., Roos, R., Maat-Kievit, A. and Tibben, A. (2004) Adverse effects of predictive testing for Huntington disease underestimated: long-term effects 7–10 years after the test. *Health Psychol.* **23**: 189–197.

Trottier, Y. and Mandel, J. (2001) Huntingtin-profit and loss. *Science* **293**: 445–446.

Van't Spijker, A. and ten Kroode, H (1997) Psychological aspects of genetic counselling: a review of the experience with Huntington's disease. *Patient Educat. Counsel.* **32**: 33–40.

Wahlin, T.B., Lundin, A., Backman, l., Almqvist, E., Jaegermark, A., Winblad, B. and Anvret, M. (1997) Reactions to predictive testing in Huntington disease: case reports of coping with a new genetic status. *Am. J. Med. Genet.* **73**: 356–365.

Went, L., Broholm, J., Cassiman, J. *et al.* (1994) Guidelines for the molecular genetics predictive test in Huntington's disease. *J. Med. Genet.* **31**(7): 555–559.

Williams, J.K., Schutte, D.L., Evers, C. and Holkup, P.A. (2000) Redefinition: coping with normal results from presymptomatic gene testing for neurodegenerative disorders. *Res. Nurs. Health* **23**: 260–269.

Williams, J.K. and Sobel, S. Dementing genetic conditions. In: *Individuals, Families, and the New Genetics* (eds S. Miller, S. McDaniel, J. Rolland and S. Feetham). (in press).

Further resources

Guttmacher, A.E., Collins, F.S. and Drazen, J.M. (2004) *Genomic Medicine: Articles from the New England Journal of Medicine*. Johns Hopkins University Press, Baltimore, OH.

Human Genome Research Institute. The Human Genome Project: Exploring Our Molecular Selves. [Accessed March 3, 2005] http://www.nhgri.nih.gov/Education/

Jorde, L.B., Carey, J.C., Bamshad, M.J. and White, R.L. (2003) *Medical Genetics*, Third Edition, St Louis, Mosby.

Nussbaum, R.L., McInnes, R.R. and Willard, H.F. (2001) *Thompson and Thompson Genetics in Medicine*. 6th Edn. W.B. Saunders, Philadelphia, PA.

Secretary's Advisory Committee on Genetic Testing (2000). A public consultation on oversight of genetic tests. [Accessed March 3, 2005] http://www4.od.nih.gov/oba/sacgt.htm

Williams, J.K., Schutte, D.L., Evers, C. and Holkup, P.A. (2000) Redefinition: coping with normal results from presymptomatic gene testing for neurodegenerative disorders. *Res. Nurs. Health* **23**: 260–269.

See 'Web-based resources for genetic healthcare' for more resources on genetic testing.

9 Autosomal recessive inheritance

9.1 Introduction to autosomal recessive inheritance

There are literally thousands of autosomal recessive conditions identified and more details of them can be found in the Online Mendelian Inheritance in Man (OMIM) resource. Because an individual with a recessive condition has no normal copy of a particular gene, the recessive conditions tend to have a serious effect on the person's health and, frequently, there is an impact on the basic body metabolism. In this chapter, the hemoglobinopathies will be used as an example of this range of conditions.

9.2 Autosomal-recessive inheritance

Genes that have an autosomal locus, such as the globin genes, will be present in two copies: one inherited from the father and one from the mother. In autosomal recessive conditions, it is necessary to inherit two copies of the variant gene to express the disease. For example, a child with sickle cell disease (SCD) will have inherited two copies of a mutation that causes SCD and both parents will be carriers.

In recessive disorders, carriers do not normally have manifestations of the disease. For any carrier couple, the chance that they will both pass on one copy of the mutated gene is 1 in 4 or 25%. As shown in Figure 9.1, each carrier parent will have a 1 in 2 or 50% chance of passing on the mutation.

Features of autosomal recessive pattern of inheritance

The features of autosomal recessive inheritance are:

- parents of affected people are normally unaffected;

Figure 9.1 Four possible outcomes from carrier parents

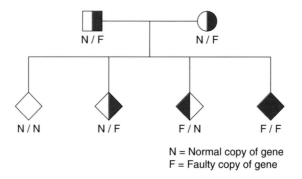

N = Normal copy of gene
F = Faulty copy of gene

- affects either sex;
- the risk to siblings of an affected child is 25%.

Calculation of genetic risks in recessive disorders

Clients may visit a genetics clinic with specific questions about the risk to themselves or their children. From knowledge about the pattern of inheritance, it is possible to work out what the various possibilities are.

Cystic fibrosis is a recessive condition that is common in northern European populations and we can use it as an example to work through what the possible consequences are.

CASE STUDY – MARK

Mark has a brother with cystic fibrosis (CF). He asks about his risk of being a carrier of the gene mutation. Mark's brother inherited a faulty copy of the CFTR (cystic fibrosis transmembrane receptor) gene from both parents. Both Mark's parents are therefore carriers. Each time they had a child, there were four possible combinations of alleles (Figure 9.1). However, Mark does not have CF, so there are three combinations remaining. In two of those combinations, a faulty gene is passed on; therefore, Mark's chance of being a carrier is two out of three, or a ⅔ chance.

Calculation of recurrence risk

The use of the Hardy–Weinberg equation to calculate population carrier rates for autosomal recessive conditions is described fully in Section 3.3. It is used here to calculate the carrier risk of the partner who does not have a family history of CF.

In this northern European population, 1 in 1600 children are born with CF.

The number of homozygotes with the recessive condition is 1 in 1600; therefore:

$q^2 = \frac{1}{1600}$

$q = $ the square root of $\frac{1}{1600}$

$q = \frac{1}{40}$

$2q = \frac{1}{20}$

The carrier risk for someone with no family history of CF in this population is 1 in 20.

CASE STUDY – MARK

Mark's risk of being a carrier of CF is ⅔. His partner Sally has no family history of CF. Her risk of being a carrier in this population is 1 in 20.

The chance that they will have a child with CF is:

Mark's risk, i.e. ⅔

Multiplied by Sally's risk, i.e. $\frac{1}{20}$

Multiplied by the chance of a child inheriting both faulty copies of the gene, i.e. ¼

$\frac{2}{3} \times \frac{1}{20} \times \frac{1}{4} = \frac{1}{120}$

9.3 The hemoglobinopathies

The molecular and protein basis of SCD and the other hemoglobinopathies is well established and, as such, they provide an illustration of the pathway between mutation and disease.

Sickle cell disease

SCD is one of a group of disorders – the hemoglobinopathies – which are caused by mutations in the genes for the globin that makes up hemoglobin. Hemoglobin is located in red blood cells and its function is the

---

CASE STUDY – MARY

Mary Walker has asked to see someone to talk about the implications of a recent diagnosis of sickle cell anemia in her son. Because of work and family commitments, she had moved during her pregnancy and had missed some of the routine antenatal care. She had also left hospital only a few hours after her son was born. From an early age he had failed to thrive, having repeated infections and episodes of painful swollen hands and feet. She had been to see her family doctor on many occasions. After some months, a blood test was ordered and the diagnosis was made. She has since seen the pediatrician to talk about her son's care and has been in touch with a support group. They have sent her a lot of information and she now wants to talk over the genetic aspects of the condition. She feels that the diagnosis of her son took a long time to make and that she was not believed when she was telling people that she knew there was something wrong with him. She is also angry that a health professional told her that her son could not have SCD but did not enquire about the ethnic background of the family.

---

absorption and release of oxygen. In SCD, the structure of the hemoglobin molecule changes under conditions of low oxygen. This causes the shape of the blood cell to change from a bi-concave disc to a crescent or sickle shape. Initially, this change is reversible when the blood is reoxygenated; however, after repeated episodes of sickling, the red blood cell is irreversibly damaged. This sickle shape leads to blockage of small blood vessels, resulting in pain and various clinical presentations. The natural history of the disease is variable.

Sickling crisis   When the red blood cells sickle, their shape changes and they become rigid and unable to transport oxygen efficiently. If the blood flow is obstructed, there is ischemia and tissue infarction. The individual experiences excruciating pain, which is often referred to as a 'sickle cell crisis'.

A number of factors may precipitate a sickle cell crisis; for example, hypoxia, acidosis, dehydration, infection, additional physical stress (e.g. physical activity or pregnancy), extreme fatigue, or trauma.

Possible complications of sickle cell disease   There are a number of possible complications, but the frequency of any individual one is very variable and some, all or none of them may occur (Gray *et al.*, 1991; Platt *et al.* 1994).

They also differ in adults and children. Possible complications include:

- vaso-occlusion;
- hemolytic anemia;
- gallstones;
- immunosuppression;
- delayed growth;
- aplastic crisis;
- splenic sequestration of blood;
- heart failure;
- leg ulcers;
- retinopathy.

The occurrence of complications depends on how effectively the disorder is managed both by health professionals and the individual; for example, the use of prophylactic antibiotics to reduce the risk of a crisis associated with infection. Some of the complications that occur in adulthood are as a result of frequent and chronic damage to tissues and organs. Therefore, the aim of treatment should be to reduce the possibility of these complications occurring.

Pathogenesis of the hemoglobinopathies

Hemoglobin is formed of two parts: heme, the iron compound which binds with the oxygen; and globins, which are proteins. Adult blood contains a mixture of hemoglobins, the most common being hemoglobin A. Hemoglobin is formed from four protein chains and two heme molecules.

In hemoglobin A, the globin chains are a pair of alpha and a pair of beta chains. These protein chains are coded for by a pair of alpha genes and a pair of beta genes. The hemoglobinopathies are autosomal-recessive disorders that are caused by mutations in these genes. SCD is caused by a point mutation that leads to an amino-acid substitution in the beta globin chain. This substitution leads to a structurally different molecule when the chains are assembled in the hemoglobin molecule (the most common being hemoglobin S). The other common group of hemoglobinopathies are the thalassemias; these are named after the chain that is deficient, alpha or beta. Thalassemias are caused by imbalances in the production of alpha and beta chains. Hundreds of mutations have been identified, but deletions of alpha genes are the most common forms of alpha thalassemia. Clinically severe conditions occur when either both beta genes or three or four alpha chains are affected. The potential combinations of variant globin chains and the relationship between the genetic variation of the protein formation and the eventual presence or absence of disease are complex. The genetics professional should have accurate and up-to-date information regarding the potential consequences of any of these combinations.

A distinction must be made between carriers who have only one affected globin locus and remain healthy, and people who have two affected globin loci (homozygous or compound heterozygous) and have one of the hemoglobinopathies. Resources regarding the clinical consequences of the various combinations of the variant hemoglobins are given at the end of this chapter in 'Further resources'.

Synthesis of globin

Hemoglobin is made up of two alpha chains and two non-alpha chains, each set of chains is associated with a heme molecule (Figure 9.2). The biophysical properties of this molecule that is formed from these four chains allow the control of oxygen uptake in the lungs and the release of oxygen in the tissues. One of the chains is designated alpha. Apart from during the very first weeks of embryogenesis, one of the globin chains is always an alpha chain. In adult

Figure 9.2 Structure of the hemoglobin molecule

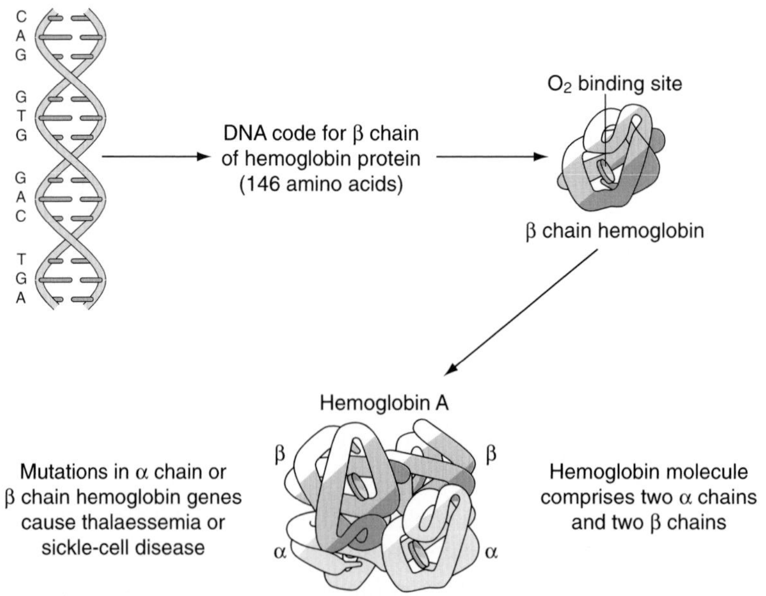

life, the second globin chain is normally a beta chain. The fetus has a specific form of hemoglobin that is formed by alpha chains and gamma chains.

With the exception of the first 10–12 weeks after conception, fetal hemoglobin is the primary hemoglobin in the developing fetus. It is termed hemoglobin F and, as discussed above, is formed from two alpha chains and two gamma chains. Hemoglobin A, the adult hemoglobin, predominates from about 18–24 weeks of age.

The genes that encode the alpha-globin chains are on chromosome 16. Those that encode the non-alpha-globin chains are on chromosome 11. There are multiple genes at each site, together with pseudo genes that are normally not expressed. The alpha complex is called the alpha-globin locus, and the non-alpha complex is called the beta-globin locus. The expression of the alpha and non-alpha genes is closely balanced by an unknown mechanism. Balanced gene expression is required for normal red-cell function.

Each chromosome 16 has two alpha-globin genes that are placed one after the other on the chromosome. For practical purposes, the two alpha-globin genes (termed *alpha1* and *alpha2*) are identical. As each cell has two copies of chromosome 16, a total of four alpha-globin genes exist in each cell. The zeta genes are expressed for a short time in early fetal life.

The genes in the beta-globin locus are arranged sequentially from the 5′ end of the chromosome to the 3′ end, beginning with the gene expressed in embryonic development (called epsilon, and expressed during the first 12 weeks after conception). The beta-globin locus ends with the adult beta-globin gene. There are two copies of the gamma gene on each chromosome 11. The others are present in single copies. Therefore, each cell has two beta-globin genes, one on each of the two copies of chromosome 11 in the cell. These two beta-globin genes express their globin protein in a quantity that precisely matches that of the four alpha-globin genes. The mechanism of this balanced expression is unknown.

9.4 Carrier frequencies

The inherited hemoglobin disorders are the most common inherited single-gene disorders. However, their prevalence has a particular geographic distribution (Davies *et al.*, 2000; Figure 9.3).

The frequency of carriers of hemoglobin S in Africa is 19–27%. Areas of the world in which carriers of mutations in genes for the hemoglobinopathies are

Figure 9.3 Distribution of the hemoglobinopathies

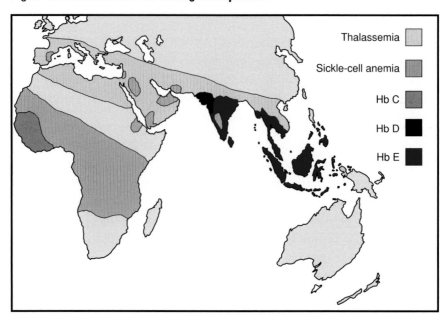

common are also areas where there is a high prevalence of malaria. There is evidence to support the suggestion that this high frequency of gene mutations for both sickle cell and alpha thalassemia is because they confer protection against malaria. In areas where malaria is endemic, carriers of SCD will be protected and will therefore survive to reproduce and pass on their genes. As this is a recessive condition and heterozygotes are more common than homozygotes, this selective advantage outweighs the fact that some homozygotes will not survive because of their disease.

Although the carrier frequency of SCD is high in individuals of African origin, it is important for health workers not to make assumptions about a person's ethnic origin. In hemoglobinopathy screening programs, in which screening has been targeted at particular ethnic groups, it has been shown that this strategy misses at least 10% of those who might have been diagnosed at birth. In addition, there are practical problems of determining race and ethnicity at birth (Gessner *et al.*, 1996; Shafer *et al.*, 1996).

CASE STUDY – MARY

When Mary and her partner attend the genetics clinic, they are still angry that the diagnosis of their son took some time to make. Mary had heard from her sister that SCD was more common in people who originated from Africa and felt that this should have been recognized in her case.

After talking through the genetics of SCD with Mary, the genetics nurse/counselor asks her if the rest of her family know about the diagnosis in her son. Mary tells the counselor that they do and she has just found out that her brother's partner Jo is pregnant. Jo had been tested as part of an antenatal screening program and knew that she also carried the sickle-cell gene. She had not wanted to tell Mary because she was thinking about having prenatal diagnosis and was worried that Mary might feel that this would mean that she should not have had her son. The counselor reflected on this when she had clinical supervision. It had brought home to her that although the genetics was essentially simple, the knowledge and implications for the family were not.

9.5 Molecular testing

In the hemoglobinopathies, biochemical testing is the first choice for diagnosis and carrier detection. It is possible to directly test for mutations, but the complexity of the number of potential mutations that might be implicated means that, at the time of publication, DNA analysis is not used as the primary testing method. In CF, the situation regarding testing for mutations is relatively more simple.

Molecular genetic laboratories use a variety of different techniques to detect mutations or to track genes through families. The basis of much genetic testing currently carried out in clinical genetics is the polymerase chain reaction (PCR). With PCR, specific DNA sequences can be copied many times to yield large quantities of the particular portion of DNA corresponding to genes or fragments of genes. These can then be analyzed or manipulated. The exact method of analysis used will depend on the characteristics of the DNA sequence of interest and the type of potential mutations. The technique of PCR has also been important in the development of forensic DNA analysis. It has allowed the amplification of small amounts of DNA to provide a unique DNA fingerprint, which can be used as evidence (Jeffreys et al., 1985).

Polymerase chain reaction

Amplification of the target DNA sequence occurs through repeated cycles of DNA synthesis. In clinical genetics, the usual source for the DNA template that is used for the reaction is DNA extracted from cells from a patient. As in karyotyping, this can be any nucleated cell. Commonly, a blood sample is used. DNA can also easily be extracted from cells in the saliva or scraped from the inside of the cheek (buccal smear). The PCR technique can also be used for prenatal diagnosis. Although DNA analysis is possible on cells harvested from amniotic fluid, the preferred sample is a chorionic villus biopsy, as this will yield greater amounts of DNA.

In addition to the target DNA, the other components of a PCR are:

- primers – short sequences of single-stranded DNA that bind by complementary base-pairing either side of the sequence containing the gene of interest;
- DNA polymerase – an enzyme that copies the target sequence and that acts at a specific temperature;
- molecules corresponding to the four bases – these are used in the synthesis of new DNA strands.

In a PCR (Figure 9.4), the template DNA (usually all the chromosomal DNA) is put in solution with the primers, the DNA polymerase and the bases. The reaction is heated, which destabilizes the double-helical structure of the template DNA. This separates into two single strands (denaturing).

The reaction is then cooled to a temperature that allows the primers to bind to the single-stranded DNA without allowing the double helix to reform (annealing). It is then heated to the temperature that allows the DNA polymerase to become active. The polymerase copies the sequence of the template DNA, starting at the primers and using the bases present in the reaction for synthesis of new single DNA strands (synthesis).

This cycle is repeated 20–40 times, depending on how much of the target sequence was present in the original DNA template. In the first cycle, synthesis carries on beyond the end of the target sequence because there is nothing to stop it. Over subsequent cycles, the newly synthesized strands, which end with a primer sequence, themselves act as templates and eventually only the target sequence is amplified.

Once sufficient quantities of target DNA have been generated, it can be used for clinical analysis or research. Techniques used to identify specific known gene mutations depend on the nature of the mutation. Analysis of specific

Figure 9.4 PCR

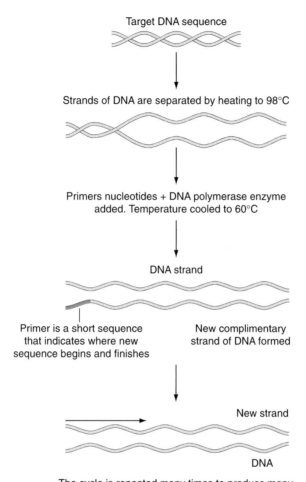

Target DNA sequence

Strands of DNA are separated by heating to 98°C

Primers nucleotides + DNA polymerase enzyme added. Temperature cooled to 60°C

DNA strand

Primer is a short sequence that indicates where new sequence begins and finishes

New complimentary strand of DNA formed

New strand

DNA

The cycle is repeated many times to produce many identical strands of the DNA sequence of interest

mutations often uses DNA probes that hybridize to the mutation. DNA probes are single-stranded specific DNA sequences; they are radioactively or fluorescently labeled. If a probe is specific for a mutation, then it will show a signal if the mutation is present and will not show a signal if it is absent. Probes may be designed to detect either normal or mutated sequences.

Southern blotting

Southern blotting (named after its inventor, Ed Southern) was the first method of analysis of DNA that utilized hybridization (Southern, 1992). The technique can be used on large fragments of DNA, which are too large to be amplified in a PCR reaction.

The DNA is cut into fragments using restriction enzymes that will only cut at specific sequences of bases (e.g. after a CTG sequence). The DNA is extracted from the cell and cut into fragments. Because the restriction enzyme cuts at specific sites, a number of different-sized fragments are produced; these are called restriction fragment-length polymorphisms (RFLPs). Their length may be altered by a mutation. The fragments are separated by size by placing the reaction on an electrophoretic gel.

Because the fragments are slightly negatively charged, they move from the negative end of the gel to the positive end. The larger fragments do not move

Figure 9.5 Southern blotting

Genomic DNA is obtained from a sample of tissue (usually blood)

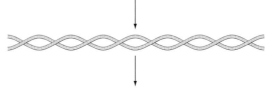

A restriction enzyme is used to cut the DNA into small fragments

The fragments are subjected to electrophoresis. Smaller fragments move further along the gel than larger ones

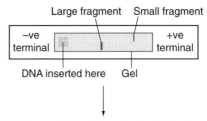

DNA fragments transferred onto a filter sheet
Filter sheet is put into a bath of radioactive probes
After the probes have attached to specific DNA fragments, the filter sheet is exposed to X-rays creating an autoradiograph or autorad

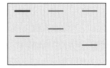

as far as the smaller fragments and therefore the larger ones are near the top of the gel and the smaller ones are near the bottom. The fragments are then transferred to a membrane by blotting. A labeled DNA probe is hybridized to the membrane. If the probe is radioactively labeled, the membrane is placed next to an X-ray file and the film is developed.

After several hours, the film will reveal a number of bands corresponding to the fragments to which the probe has hybridized. A specific mutation can be detected if it alters a restriction enzyme cutting site, so that the presence of a mutation causes a different band pattern to that caused by the absence of a mutation. Large deletions or insertions will also change the length of the restriction fragments.

Southern blotting has also been used extensively in research to isolate genes and to track them through families. Southern blotting analyzes DNA. Another similar technique – called northern blotting – is used to analyze RNA.

Testing for mutations

A direct test for a mutation can be performed if:

- a specific mutation is known;
- there is a specific probe for either its presence or absence;
- the mutation alters a restriction enzyme cutting site or the size of a band.

In the case of CF, more than 1000 different mutations have been reported in the *CFTR* gene, but most of these are very rare. The most common CF mutation is DF508 (also called Delta F508 or ΔF508), which is a three-base-pair deletion in the gene. This mutation accounts for about 70–75% of CF cases occuring in individuals originating from northern Europe, although there is some regional variation. The frequency of this mutation also varies in different ethnic groups, accounting for less than 50% of cases in Jewish, Afro–Caribbean, and Asian populations. The frequency of some other relatively common mutations is around 1–3%, but some mutations are so rare that they are only known to exist in individual families.

Laboratories do not routinely check for all known mutations. If the mutation has been identified in a family, accurate carrier testing will be available for that family. If it has not, then the information provided by a negative test will depend on the probability of the mutation being causative in the particular population in previous years. The ΔF508 mutation is common enough in the northern European population to be useful for testing in the general population, but the situation might be different for someone from a different ethnic group.

9.6 Finding disease genes

The story of the discovery of the gene that confers susceptibility to hemochromatosis illustrates the complexity of finding genes and also highlights the use of a number of molecular genetic techniques. In SCD, the variant protein was known before the gene had been identified. The genes were identified through 'functional cloning', working from the protein back to the gene. In hemochromatosis, there was no clue as to the underlying protein abnormality until the gene was identified.

Hemochromatosis is an autosomal-recessive disorder of iron metabolism. In patients affected with hemochromatosis, the regulation of iron absorption is disturbed. Iron is essential for the body; however, it is also toxic in excess. Iron is not excreted and therefore control is at the level of the gut, where absorption of iron is upregulated in conditions of iron deficiency and downregulated when body stores are sufficient. Individuals who have hemochromatosis continue to absorb iron even when they are iron-loaded. This excess iron is stored in the liver and other organs, and can eventually lead to these organs being damaged.

The first step in the identification of the gene for hemochromatosis was taken using tissue typing and the human leukocyte antigen (HLA) locus. This locus, on chromosome 6, shows great variability between individuals as it is associated with immunity. Tissue typing for transplantation is based on characterizing the HLA type. Before the development of molecular genetics, the variation in HLA types could be used to look for linkage between particular HLA types and diseases.

In 1976, Simon found an excess of HLA-A3 alleles among individuals with hemochromatosis, revealing that hemochromatosis is inherited as an autosomal-recessive trait that is caused by a gene tightly linked to the HLA-A locus on chromosome 6 (Simon *et al.*, 1976). The cloning of the *HHC* gene was made possible by a number of advances in genetics in the late 1980s. Location of the gene had been mapped to a relatively small section of chromosome 6. Linkage analysis is covered in more detail in Chapter 10.

Because the position of the gene was known. a positional cloning strategy could be developed (Figure 9.6). If the location but not the nature of the gene

Figure 9.6 Functional and positional cloning

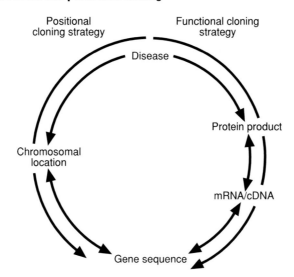

is known, the region is mapped and genes in the region are characterized, allowing the protein and its defects to be identified. However, the genetic distance was considerable and was estimated – on the basis of the number of recombination events occurring between these loci – to be several centimorgans or megabases.

In the early 1990s, the magnitude of the task of analyzing such a stretch of DNA was little short of that facing the Human Genome Project, which was in its infancy. There was also an absence of a map for this region of chromosome 6. Therefore, a number of groups set out to construct one. At this time, DNA maps were constructed by analyzing small sections of overlapping DNA and using a variety of techniques to assign their position on the chromosome.

Various groups narrowed the candidate region and eventually, in 1996, Feder and colleagues carried out a study in 101 hemochromatosis patients and 64 controls, using 45 of the DNA markers contained within the candidate region. This group genotyped or 'finger-printed' them by determining the combination of alleles present in each individual. Once the region was reduced to 250 kilobases, the DNA contained in that region could be sequenced and two mutations were identified that appeared to be associated with hemochromatosis. This enabled the *HFE* gene to be identified. Further studies in animals and humans were required to confirm that the *HFE* gene was associated with hemochromatosis.

Feder's study identified a gene in which two missense mutations accounted for 88% of the affected probands. This gene – previously called *HLA-H* – is now called *HFE*. Further prevalence studies have confirmed the relationship between the two common mutations and hemochromatosis. These mutations are a G to A transition at nucleotide 845 on the HFE gene, causing aspartate to substitute for histidine at position 282 on the HFE protein (C282Y), and a G to C transition at nucleotide 187, causing an aspartate to histidine substitution at position 63 in the HFE protein (H63D). The C282Y mutation disrupts the structure of the protein and means that it is not able to be transported to the cell surface. The mechanism by which the H63D mutation leads to excess iron absorption is not clear.

The majority of patients are homozygous for the C282Y mutation, with a smaller minority being compound heterozygotes for both mutations (The UK Haemochromatosis Consortium, 1997). In a recent review of genotyping studies in clinically affected probands, the frequency of homozygosity for C282Y ranged from 52–100%, with approximately 5% of cases being compound heterozygotes for the two mutations (Hanson *et al.*, 2001). This review included studies that were from ethnically diverse populations, some of whom would not be expected to have a high frequency of the C282Y mutation.

As many as 10% of cases of hemochromatosis are not accounted for by these common mutations. Other mutations within the *HFE* gene have been identified, but their clinical significance is not yet clear (Pointon *et al.*, 2000). In addition, rare cases have been reported in families that have iron-overload disorders that have been shown to be caused by mutations in genes for other proteins involved in iron-metabolism pathways.

Chromosome-6-linked hemochromatosis is generally considered to be a disease that occurs in Europeans. Prevalence studies have, indeed, indicated that the common mutations are found primarily in individuals of northern European descent, with particularly high prevalence in 'Celtic' populations (Merryweather-Clarke *et al.*, 2000).

Unlike the heterozygote advantage of resistance to malaria conferred by hemoglobinopathy genes, the explanation for the high frequency of mutations in the *HFE* gene is not known. It is hypothesized that possession of an *HFE* allele will be protective against iron-deficiency anemia, which might lead to a reproductive advantage.

The success of positional cloning in identifying the HFE protein led to the identification of further proteins involved in iron metabolism. However, the function of the HFE protein and its role in the regulation of iron metabolism is still not fully clear.

9.7 Population screening

CASE STUDY – MARY

Michael and Jo, Mary's brother and his partner, come to the genetics clinic for Michael to be tested. Jo had been offered a test to see if she was a carrier of SCD when she went for her first antenatal check-up. She had been found to be a carrier and wanted to talk about what they should do about it. She was relieved to hear that the baby would only have a risk of having SCD if Michael was also a carrier. His blood test was arranged. The counselor encouraged Jo and Michael to explore what they would do with the information if Michael was a carrier and outlined their options: to do nothing, to have prenatal diagnosis simply for information and to be prepared, or to have prenatal diagnosis and consider terminating the pregnancy if the baby was shown to have SCD. The test on Michael showed that he was not a carrier and, to their great relief, they continued with the pregnancy.

The above is an example of screening to detect carriers of a recessive condition in order to offer reproductive choice. More commonly, screening would be to detect a child or an adult who has a disorder in order to intervene in some way to alter the consequences of that disorder. It is important to realize that there is a difference between testing and screening.

The first report of the UK National Screening Committee (NSC, 1988) states that 'Screening is a public health service in which members of a defined population, who do not necessarily perceive they are at risk of, or are already affected by a disease or its complications, are asked a question or offered a test, to identify those individuals who are more likely to be helped than harmed by further tests or treatment to reduce the risk of a disease or its complications.' Genetic testing is usually targeted at a group for whom the risk of the condition is significant due to family history or other indicators.

The primary objective in evaluating screening programs should be to balance the benefits and harms of the program, the benefit being whether the program reduces the risk of morbidity or mortality of the disease that is being screened for. The evaluation should include both the benefit that is produced in terms of risk reduction balanced against the potential harms caused by screening. This definition has to be modified to some extent when considering antenatal screening programs – for example, the purpose of screening for Down syndrome is to allow pregnant couples to make an informed choice.

As screening involves approaching a large number of individuals to detect a much smaller number of individuals who might benefit from the intervention, the assessment of potential harm caused to those who screen negative is particularly relevant when screening programs are being evaluated. The classic criteria for evaluation of screening programs (suggested by Wilson and Jungner, 1968) establish the importance of considering a number of factors before recommending the implementation of a screening program. In summary, these are that:

- the condition should be an important health problem;
- the natural history and epidemiology should be understood;
- there should be a recognized pre-symptomatic or latent period;
- the test should be acceptable, safe and reliable;
- effective and acceptable treatment should be available;
- all other options for managing the condition should have been considered.

These criteria have been expanded and there is now an international consensus that they should take account of the more rigorous standards that are needed to demonstrate effectiveness and greater concern about possible harm caused by screening. In particular, it is proposed that there should be 'evidence from high-quality randomized controlled trials that the screening program is effective at reducing mortality or morbidity'; evidence that the complete program is clinically, socially and ethically acceptable; evidence that the benefit outweighs the harm and that all other options for managing the condition have been considered.

Neonatal screening

Newborn screening for phenylketonuria (PKU) is a paradigm for the success of neonatal screening in that the condition is serious, there is effective treatment, and agreement on the screening test and screening is the most effective way to identify and treat those who have the condition. Once a child has enough symptoms to be diagnosed, there will be irreversible damage. Although PKU is a genetic disease, the range of mutations that are responsible means that a DNA-based diagnosis is currently not as useful as a direct biochemical measurement.

PKU is an autosomally recessive inherited inborn error of metabolism in which affected individuals are unable to metabolize the amino acid phenylalanine to tyrosine. This leads to high levels of phenylalanine that are neurotoxic. In the absence of treatment, nearly all affected individuals develop severe, irreversible learning difficulties, together with neurological deterioration. These severe manifestations are now very rarely seen due to universal neonatal screening and subsequent dietary restriction. Classical PKU is caused by a deficiency of the enzyme phenylalanine hydroxylase (PAH); varying degrees of PAH deficiency, as well as disorders of other enzymes in the metabolic pathway, can all lead to high levels of phenylalanine (Fig 9.7). The disorder is therefore 'heterogeneous', or caused by one of a number of possible genetic defects.

All babies should have their phenylalanine levels measured on a dried blood spot. This should be done 6–14 days after birth. Early diagnosis allows the initiation of a diet that is low in phenylalanine. In effect, this means that protein intake is extremely restricted and that supplementary amino acids are taken. Recent data indicate that this diet should continue at least until adulthood. There are special concerns about pregnancy in women who are affected with PKU and it is essential that low serum phenylalanine levels are maintained before conception and throughout the pregnancy. Although PKU is relatively rare (1 in 12 000 births), the prognosis for the child if it is untreated is devastating. The fact that there is an effective treatment and a screening test with high sensitivity and specificity make compelling arguments for universal population screening.

Neonatal screening for SCD is offered widely in the USA (National Newborn Screening and Genetics Resource Center, 2004) and is being developed to be

Figure 9.7 Enzyme pathway involved in phenylketonuria

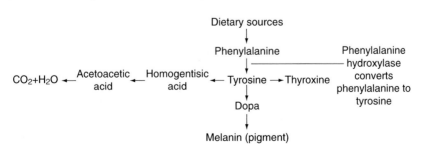

offered in the UK. The benefit is in the management of affected children, primarily by the use of prophylactic antibiotics. The evidence has been reviewed in documents produced by organizations concerned with public-health initiatives.

In the UK. a decision has been made not to offer population screening for medium-chain acetyl-CoA dehydrogenase deficiency (MCADD), although it has been considered. This is a condition in which the child may suffer severe hypoglycemia, encephalopathy, or sudden death as an abnormal response to low food intake or dehydration. In diagnosed children, it is possible to avoid these adverse consequences by dietary management. The National Screening Committee has suggested that it is premature to introduce screening for this condition as it is not clear how many individuals with the abnormality have a risk of the extreme consequences and there are concerns about the performance of the screening test. This decision is subject to regular review. In this classical approach to screening, the benefit that is assumed is to the child and it is agreed that there should be evidence that the benefit outweighs potential harm before screening is introduced. However, a different decision has been reached in many of the states in the USA, where it is offered as part of screening for multiple inborn errors of metabolism using mass tandem spectroscopy.

The March of Dimes (a US organization that aims to improve the health of babies) has recently suggested that all newborns in the USA should be screened for nine inborn errors of metabolism, most of which are recessive conditions (March of Dimes, 2004). In the UK, newborn screening is always voluntary. There is no mandatory testing, although until recently this has been on the basis of limited information being given to parents. This situation is different in other countries, although it is increasingly the case that informed consent for screening is seen to be important (American Academy of Pediatrics, 2001).

Some of the newborn screening tests will identify carriers of the condition that are being screened for. This raises concerns about the nature of information that should be passed on and the potential for abuse of that information. Early SCD screening programs that were not preceded by appropriate community education did lead to misunderstandings, discrimination and stigmatization of carriers. It is generally accepted that screening in the newborn period for the purpose of detecting carriers is not appropriate. It is important that, when planning screening programs, attention is paid to how to manage the carrier information that is generated as a result of the testing process (American Academy of Pediatrics, 2001; Laird *et al.*, 1996).

Hemochromatosis

Decisions about introducing screening programs require evaluation of the evidence about the benefits and harms of introducing such programs, and different decision-makers may come to different conclusions depending on local circumstances. There is a consensus, however, regarding population screening for hemochromatosis and this has been reached following careful evaluation of the evidence.

As the predisposition to hemochromatosis is common and there are presymptomatic tests and treatment, population screening has been considered as a possible strategy for early case identification in this condition. The College of American Pathologists recommended population screening using transferrin saturation level in 1996 (Witte *et al.*, 1996) and several cost-effectiveness studies have suggested that it might be an appropriate strategy.

The identification of the gene led to calls for the implementation of genetic screening programs (Allen and Williamson, 2000). However, several authors, particularly those working in the public health field, advised caution and more careful evaluation before implementation of screening strategies that utilize either genetic or biochemical testing (Haddow and Bradley, 1999). It was assumed that the mutation conferred a high risk of developing iron overload. However, further studies in representative populations suggested that the risk of developing clinically important iron overload in individuals who had this mutation was possibly as low as 1% (Beutler *et al.*, 2002).

The predictive value of this mutation test for population screening is probably not clinically useful. The patients in the original studies represent the severe end of the spectrum, and other genetic and environmental factors modify risk. In addition, as a result of this, more careful evaluation of the consequences of abnormal iron indices are also being questioned. However, hemochromatosis has preventable consequences for those who have it, and first-degree family members who are at risk should be offered testing.

9.8 Conclusion

Recessive genetic conditions are a common reason for referral to clinical genetics services. There are various issues for families, including reproductive decision-making, family testing or diagnosis. In the family context, the genetics health professional can use their skills in risk estimation to inform families and provide a framework for decision-making. Because certain recessive conditions have been well-characterized, they have been used in this chapter to illustrate the discovery of genes and also the pathway from gene to disease. Certain recessive conditions have proved amenable to population screening and the extension of genetic practice from families to the population requires an understanding of the principles of screening in order to maximize the benefits and minimize the harms of this approach to healthcare.

Study questions

Case scenario – Alan Jones

Alan Jones rings up the department. His brother has an inoperable liver cancer (he is 42 years old) and has been told that the cancer was caused by hereditary hemochromatosis. Alan has been on the internet and read the information from an international hemochromatosis society. He knows that the complications of hemochromatosis, if diagnosed early enough, are preventable.

Task A
(a) What information do you need from Alan?
(b) What is the mode of inheritance of hemochromatosis?
(c) What are the clinical consequences of hemochromatosis?
(d) What risk would you give Alan?
(c) Would you offer Alan a test, and what would you test for?

Task B
You meet Alan and his wife, Julie, in the clinic. Unfortunately, Alan's brother has died while on the waiting list for a liver transplant. Alan has read a great deal about the condition. He tells you that he must have it because he is constantly tired, and suffers with pain in his back and knees. Julie is worried about Alan and their children, a boy and a girl, but is also very anxious to make sure that her two nieces are tested for this condition.

(a) What would you say to Alan about his certainty that he does have hemochromatosis?
(b) What would you say about testing Alan and Julie's children?

(c) What information would you give Alan about his results?

Task C

Later in the year, Sarah, Alan's first cousin, is referred to the department. Would you offer Sarah a test?

References

Allen, K. and Williamson, R. (2000) Screening for hereditary haemochromatosis should be implemented now. *Br. Med. J.* **320:** 183–184.

American Academy of Pediatrics (2001) Ethical issues with genetic testing in pediatrics. *Pediatrics* **107:** 1451–1455.

Beutler, E., Felitti, V.J., Koziol, J.A., Ho, N.J. and Gelbart, T. (2002) Penetrance of 845G—> A (C282Y) HFE hereditary haemochromatosis mutation in the USA. *Lancet* **359:** 211–218.

Davies, S.C., Cronin, E., Gill, M., Greengross, P., Hickman, M. and Normand, C. (2000) Screening for sickle cell diseases and thalassaemia: a systematic review with supplemantery research. *Health Technol. Assess.* vol 4, no. 3.

Gessner, B.D., Teutsch, S.M. and Shaffer, P.A. (1996) A cost-effectiveness evaluation of newborn hemoglobinopathy screening from the perspective of state health care systems. *Early Hum. Dev.* **45:** 257–275.

Gray, A., Anionwu, E.N., Davies, S.C. and Brozovic, M. (1991) Patterns of mortality in sickle cell disease in the United Kingdom. *J. Clin. Pathol.* **44:** 459–463.

Haddow, J.E. and Bradley, L.A. (1999) Hereditary haemochromatosis: to screen or not – conditions for screening are not yet fulfilled. *Br. Med. J.* **319:** 531–532.

Hanson, E.H., Imperatore, G. and Burke, W. (2001) HFE gene and hereditary hemochromatosis: a HuGE review. Human Genome Epidemiology. *Am. J. Epidemiol.* **154:** 193–206.

Jeffreys, A.J., Wilson, V. and Thein, S.L. (1985) Individual-specific 'fingerprints' of human DNA. *Nature* **316:** 76–79.

Laird, L., Dezateux, C. and Anionwu, E.N. (1996) Neonatal screening for sickle cell disorders: what about the carrier infants? *BMJ* **313:** 407–411.

March of Dimes. Newborn screening recommendations. [Accessed July 28, 2004] www.marchofdimes.com/professionals/580_4043.asp

Merryweather-Clarke, A.T., Pointon, J.J., Jouanolle, A.M., Rochette, J. and Robson, K.J. (2000) Geography of HFE C282Y and H63D mutations. *Genet. Test.* **4:** 183–198.

National Newborn Screening and Genetics Resource Center (2004) Current newborn screening conditions by state. [Accessed August 2, 2004] http://genes-r-us.uthscsa.edu

NSC (1988) First Report of the NSC, Department of Health. [Accessed March 3, 2005] http://www.doh.gov.uk/nsc/

Platt, O.S., Brambilla, D.J., Rosse, W.F., Milner, P.F., Castro, O., Steinberg, M.H. and Klug, P.P. (1994) Mortality in sickle cell disease. Life expectancy and risk factors for early death. *N. Engl. J. Med.* **330:** 1639–1644.

Pointon, J.J., Wallace, D., Merryweather-Clarke, A.T. and Robson, K.J. (2000) Uncommon mutations and polymorphisms in the hemochromatosis gene. *Genet. Test.* **4:** 151–161.

Shafer, F.E., Lorey, F., Cunningham, G.C., Klumpp, C., Vichinsky, E. and Lubin, B. (1996) Newborn screening for sickle cell disease: 4 years of experience from California's newborn screening program. *J. Pediatr. Hematol. Oncol.* **18:** 36–41.

Simon, M., Bourel, M., Fauchet, R. and Genetet, B. (1976) Association of HLA A3 and HLA B14 antigens with idiopathic hemochromatosis. *Gut* **17:** 332–334.

Southern, E.M. (1992) Detection of specific sequences among DNA fragments separated by gel electrophoresis. 1975. *Biotechnology* **24:** 122–139.

The UK Haemochromatosis Consortium (1997) A simple genetic test identifies 90% of UK patients with haemochromatosis. *Gut* **41:** 841–856.

Wilson, J. and Jungner, G. (1968) *Principles and practice of screening for disease.* World Health Organization, Geneva.

Witte, D.L., Crosby, W.H., Edwards, C.Q., Fairbanks, V.F. and Mitros, F.A. (1996) Practice guideline development task force of the College of American Pathologists. Hereditary hemochromatosis. *Clin. Chim. Acta* **245:** 139–200.

Further resources

Centre for Disease Control Genomics and Disease Prevention. [Accessed March 3, 2005] http://www.cdc.gov.genomics/activities/ogdp/2003.htm

National Newborn Screening and Genetics Resource Center (2004) Current newborn screening conditions by state. [Accessed August 2, 2004] http://genes-r-us.uthscsa.edu

National Electronic Library for Health. [Accessed March 3, 2005] http://www.nelh.nhs.uk/screening/dssp/home.htm

Online Mendelian Inheritance in Man (OMIM). [Accessed April 11, 2005] http://www.ncbi.nlm.nih.gov/entrez/query.fcgi?db=omim

10 X-linked inheritance

10.1 Introduction

The material in this chapter concerns those conditions that are caused by a mutation in a gene on the X chromosome. The majority of these conditions are recessive, but there are also X-linked conditions that follow a dominant pattern of inheritance. The phenomenon of X-inactivation (section 10.7) can make the counseling for X-linked recessive conditions complex because of the variability of effect in the carrier females.

CASE STUDY – JIM

Jim Turner was referred to a genetics service with his wife. They were planning to start a family and were concerned because Jim had a brother with fragile X syndrome. He wanted to know the possibility of him having a child with similar problems to his brother Simon. The genetics nurse counselor asked Jim what Simon was like. He was a much-loved member of the family with a great sense of humor. However, it was unlikely that he would ever be able to live independently and Jim remembered, when they were small, how difficult life had been at times, dealing with Simon's obsessions and problems in coping with changes to his routine. Jim needed to know what the chances of having a child with similar problems to Simon was, in order that he and his partner could prepare for it.

10.2 Fragile X syndrome

Fragile X syndrome is an X-linked disorder that causes mental retardation/learning difficulties that can range from mild to more severe in affected males and may also cause problems in affected females. Males who have the disorder show developmental delay, particularly in speech, but have only mild dysmorphic features: a long face, prominent jaw and large ears. The behavioral phenotype includes hyperactivity and autistic features, such as gaze aversion and shyness.

The disorder is caused by a mutation in the FMR1 gene. The mutation in more than 99% of affected individuals is caused by an increased number of CGG trinucleotide repeats (usually more than 200; Verkerk *et al.*, 1991). This mutation leads to an abnormal methylation pattern of the gene. Methylation is the process by which a gene can either be silenced or switched on. In fragile X syndrome, the gene is silenced.

The syndrome was named fragile X syndrome because the abnormal gene creates a so-called fragile site at Xq27, which can be identified on a specially prepared karyotype. Before gene cloning was developed, the method of diagnosis was via karyotyping and looking for the fragile site.

Since the isolation of the gene in 1991, prevalence estimates of the syndrome have been revised downwards (Turner *et al.*, 1996). This is partly due to the fact that cytogenetic analysis has identified other fragile sites close to the F.

10.3 X-linked inheritance

The inheritance of fragile X syndrome is complicated by unusual features of the gene, which will be discussed later in this chapter. However, the basic pattern of inheritance is X-linked.

In X-linked recessive disorders, the mutated gene is on the X chromosome. Genes on the X chromosome are present in one copy in males and two copies in females. Therefore, if there is a gene mutation, males who inherit the mutated X chromosome from their mother will be hemizygous for the mutation. Apart from in rare circumstances, females who inherit the mutated X chromosome from their mother will inherit a normal X chromosome from their father and therefore will be heterozygous.

Features of X-linked recessive disorders

The features of X-linked recessive disorders are listed here and are also represented diagrammatically in Figure 10.1:

- the genetic mutation is within a gene on the X chromosome;
- affects mainly males (occasionally females will be affected);
- the mother of an affected male will usually be an asymptomatic carrier (show no or minimal signs of the condition) but may have affected male relatives;

Figure 10.1 Inheritance risks in X-linked condition with carrier mother

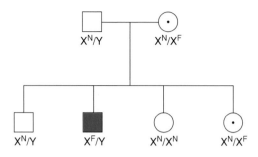

X^N = Chromosome containing normal gene
X^F = Chromosome containing faulty gene

- no male to male transmission is seen on the family tree;
- all the daughters of affected males will be carriers;
- all sons of an affected male will be unaffected.

10.4 Duchenne muscular dystrophy

Duchenne muscular dystrophy (DMD) is caused by mutations in the dystrophin gene. Becker muscular dystrophy – a milder condition – is also caused by mutations in the dystrophin gene. However, the causative mutations in the DMD gene are usually frameshift mutations (Boland *et al*, 1996). That is, they disrupt the reading frame of the gene, leading to a truncated or abnormal protein.

10.5 Mutations and mutation analysis

Disease-causing mutations usually affect the protein sequence that is produced by the gene and can be categorized in the following way.

Single base-pair substitutions or point mutations

As each protein may be coded for by several different combinations of base pairs (i.e. the code is redundant), single base-pair substitutions may change the triplet code but produce the same amino acid. For example, leucine may be encoded by CUC or CUA. A change in the last base from C to A would therefore be neutral.

As discussed previously, the code for each amino acid in the protein string is composed of three base pairs called a codon. A substitution may change the codon to a STOP codon, which would stop translation and lead to production of the protein being terminated prematurely. This is commonly called a nonsense mutation and would have a significant effect on the final protein structure.

A missense mutation occurs when the substitution and subsequent altered codon specifies a different amino acid. So, it can be seen that some substitutions will have little effect on the final protein structure, whereas others will have major effects.

Insertions and deletions

Insertion or deletion of one or more bases may also cause specific phenotypes. The effect of insertions or deletions depends on whether the number of bases involved is a multiple of three or not. As each amino acid is coded by a multiple of three bases, the way in which the DNA is read remains constant, although there may be one missing or one extra amino acid. The way this is normally referred to is that the reading frame of the DNA sequence remains constant and the mutation is said to be 'in-frame'. The exact phenotypic effects of these in-frame mutations will depend on the effect of the mutation on the structure of the encoded protein (i.e. whether the alteration seriously disrupts the final protein or not) (Figure 10.2).

If the numbers of bases inserted or deleted are not multiples of three, then the reading frame is disrupted and the sequence of amino acids downstream from the mutation will be read differently. These frameshift mutations usually have a serious effect on the eventual protein (Figure 10.4).

Figure 10.2 Types of mutation

CUU AUG CAG GGA CGG	Original code
CUU AUG GGA CGG	In-frame deletion
CUU AUG CAG **AAA** GGA CGG	In-frame insertion
CUU UGC AGG GAC GG	Frameshift deletion
CUU A**U**U GCA GGG ACG G	Frameshift insertion

The dystrophin gene is located at Xp21 and is the largest gene so far described. Gene deletions/duplications are found in ~65% of patients with DMD. Point mutations account for a substantial proportion of the remainder. Deletions are more common in specific areas of the gene and screening these mutation hotspots identifies ~98% of deletions.

Testing for mutations

There are now many different techniques that test for mutations and new methodologies are constantly being developed. In this section, a selection of the more commonly used methods are described.

Southern blotting   This has been described previously, in section 9.5, and was one of the earlier methods available for detecting mutations.

Pulse-field gel electrophoresis (PFGE)   The agarose gel used in standard gel electrophoresis, as described in Southern blotting, acts as a filter and can only allow analysis of a certain size of fragment. Fragments that are larger than 40 kb cannot move easily through the gel and therefore cannot be resolved. In PFGE, the DNA is isolated from the cells in such a way that breakage is minimized. Enzymes that cut at rare restriction sites are used. The PFGE apparatus applies a current to the gel as before; however, the current is alternated so that it periodically switches direction. The DNA fragments are therefore forced to change their shape and direction of travel. The time taken to do this is dependent on size and this allows the fragments to be separated.

Microarrays   Microarrays are a recently developed technology that mean that many different genes can be analyzed simultaneously and are the basis of the so-called 'gene chips' that are frequently talked about in the media. The technology is based on libraries of cloned DNA, which are placed on a glass slide. These are the probes and, as before, they are used to interrogate target DNA about which information is sought. The target DNA is labeled with a fluorescent probe and allowed to come into contact with the microarray, enabling probe-target complexes or heteroduplexes to form. Following hybridization, the fluorescent label is detected using a laser scanner and digital imaging software. This technology is a powerful tool that is currently used mostly in research settings to investigate RNA expression levels or for DNA variation screening.

In DMD, a variety of techniques have been developed to look for mutations. Deletion screening of the mutation hotspots is routinely available in most genetics laboratories. Testing methodology may include Multiplex PCR analysis (testing multiple mutations in one polymerase chain reaction (PCR)) or Southern blotting. Detection of point mutations is available from specialized laboratories and may only be available as part of a research program.

10.6 Carrier testing

In X-linked conditions, females carry the gene mutation and males are affected with the condition. As many X-linked conditions are serious, women may want to know their carrier status in order to make reproductive choices.

Bayesian calculations in X-linked conditions

The method for the Bayesian calculation is described in section 3.3 and is frequently used in refining the carrier risk for females in families that are affected by X-linked conditions.

CASE STUDY – SARAH

Sarah was referred to find out if she was a carrier of the gene for DMD (see Figure 10.3). Her brother, Roy, had died from the condition some years ago. Sarah was informed that, at the present time, because a mutation had not been identified in Roy, there was no direct mutation test that would say if she was a carrier of the condition. Roy's stored DNA sample was retrieved and sent to the laboratory for further investigation. It was possible to work out the risk of Sarah being a carrier based on the information that was available, using principles derived from Bayes' theorem.

Sarah gave the following family tree. There were no other affected boys in the family and, in addition to Roy, Sarah had one other older brother who was well.

The condition is X-linked; therefore, if Sarah's mother was a carrier, Sarah would have a 50% chance of carrying the gene mutation. However, it is known that a proportion (approximately one-third) of mutations in the DMD gene occur *de novo*. As there is no other family history, there is a one-third chance that this was a new mutation in the family and a two-thirds chance that Sarah's mother carries the mutation.

The prior probability that Sarah's mother is a carrier is therefore ⅔ (⅓ chance she is not a carrier). As Sarah can only be a carrier if her mother is also a carrier (barring unusual situations involving gonadal mosaicism, which will be discussed later), the counselor calculates the probability that Sarah's mother carries the mutation. The other information that is available is the fact that Sarah has an unaffected brother. Using this information, the Bayesian calculation can be done using the guide in section 3.3.

Calculation for Sarah's mother's status

	Carrier	Not a carrier
Prior probability	⅔	⅓
Probability of having a healthy son (conditional)	½	1
Joint probability	⅓	⅓

Probability that Sarah's mother is a carrier 1:1 or 50%.

Probability that Sarah is a carrier = ¼ or 25%

There is an additional piece of information that can be introduced to refine the estimate of Sarah's carrier risk further. Affected boys with DMD have highly raised levels of an enzyme called creatinine kinase. Some women who carry the gene also have raised levels of this enzyme, but this is not specific enough as a carrier test. Sarah has a blood test that shows slightly raised levels of this enzyme, and previous studies have shown that 20% of women who are carriers have levels in this range.

From the previous calculation, the chance of Sarah being a carrier has been estimated at ¼ or 25%. Using this as the prior probability, a further calculation can be performed using the information from the creatinine kinase levels.

Sarah	Carrier	Not carrier
Prior probability	¼	¾
Conditional probability	⅕	⅘
Joint probability	¹⁄₂₀	¹²⁄₂₀

Probability that Sarah is a carrier 1:12 or 8.3%.

Her chance of having an affected son is therefore 8.3% × 25% = approximately 2%.

This has lowered Sarah's carrier risk; however, she still wishes to know if there is anything else that could be offered to her. As she has an unaffected brother, and DNA samples are available from her mother, father and affected brother, it would be possible to offer Sarah further information by performing a linkage study on the family.

Linkage

Within the dystrophin gene, there are a number of highly variable (polymorphic) sequences of DNA. These variable markers can be used to track the gene through the family. Although this process does not provide any more information about the mutation, it does mean that Sarah can be told if she shares an X chromosome with either of her brothers. Linkage analysis is a

Figure 10.3 Sarah's family tree

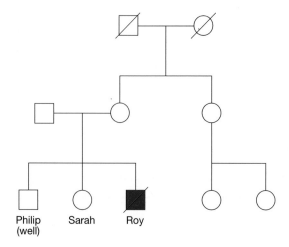

technique that has been used to successfully map and subsequently to identify many genes.

The results of the linkage analysis (Figure 10.4) show that Sarah and her two brothers seem to have inherited the same X chromosome from their mother. This indicates that Sarah's mother is not a carrier and that the mutation arose *de novo* in Michael. However, these results must be interpreted with caution.

Figure 10.4 Linkage result on Sarah's family

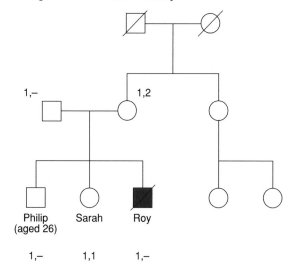

The DMD gene is so large that it is possible that a recombination could occur within the gene. As discussed in Chapter 2, recombination and exchange of genetic material is part of the process of meiosis. If a recombination event had occurred, it would change the interpretation of the results. Further analysis of the samples with more markers show that a recombination has occurred and this makes the results uninformative for determining Sarah's carrier status (Figure 10.5).

Gonadal mosaicism

In the above situation, even though the mother does not seem to carry the mutation it is not possible to reassure her daughter completely. The reason for

Figure 10.5 Recombination

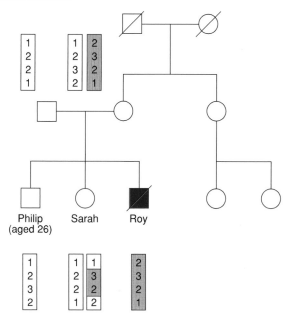

CASE STUDY – SARAH

Further analysis of Roy's sample successfully identifies the mutation and Sarah is able to have a test to determine if she is a carrier. The test shows that she is not. Sarah is relieved and also pleased that her mother informed her of the risk before she had made a decision about starting a family. The extra analysis had taken some months, which would not have been available to her if she had been pregnant. Further testing on Sarah's mother showed that she had no detectable mutation in her blood.

this being that the phenomenon of gonadal mosaicism has been well-described in DMD (Covone *et al.*, 1991). In gonadal mosaicism, a mutation arises in the germ cells, resulting in a population of egg or sperm cells that have the mutation. This confers a recurrence risk for future offspring and should be taken into account when an individual is having genetics counseling.

10.7 Females with X-linked disorders

As discussed in the previous example, some female carriers of DMD show raised levels of creatinine kinase. In addition, about 5–10% show some degree of muscle weakness and there have been reports of cardiomyopathy developing (Emery, 2002; Grain *et al.*, 2001).

Females may be affected by X-linked recessive disorders if they have a chromosome rearrangement that either disrupts the wild-type gene or, as in the case of Turner syndrome, only have one X chromosome. For conditions in which affected males can reproduce, all their daughters will be carriers (e.g. hemophilia). In this situation, if the girl's mother also carries a mutation for hemophilia, the girl would be affected.

Another explanation for manifesting carriers relates to the process of X-inactivation.

X-inactivation

In males and females, the dosage (amount) of genes is the same except for the genes carried on the sex chromosomes. There are some genes that are common to both the X and Y chromosomes, but the majority of genes on the X chromosome do not occur elsewhere. In effect, females have twice the dosage of genes on the X chromosome compared to males. To counteract this effect, a process known as X-inactivation occurs early in embryonic life. In each cell in the female embryo, one of the two X chromosomes is 'switched off,' leaving only one active X chromosome in each cell. The initiation of X-inactivation is controlled by the *XIST* gene on the X chromosome, but other genes are known to affect the function of XIST (Heard *et al.*, 1999).

The X-inactivation should occur randomly, with 50% of each of the maternally and paternally derived X chromosomes being inactivated throughout the body. The inactivation of the X chromosome is achieved by methylation, a chemical process whereby methyl groups are added to the DNA molecule at the site that is to be inactivated.

If a female is a carrier of an X-linked condition, such as hemophilia or DMD, the X chromosome that has the normal copy of the gene may be preferentially activated in most cells, leading to a skewing of the X-inactivation pattern.

X-linked dominant inheritance

There are rare conditions in which the X-linked gene seems to follow a dominant pattern of inheritance. For example, in vitamin-D-resistant rickets, females as well as males in the family are affected, although females may be less severely affected than males (Scriver *et al.*, 1991). In other conditions, such as incontinentia pigmenti, it is assumed that the mutation is lethal in males, leading to an excess of females in the families. In this condition, the mutation causes abnormalities of the skin, hair, and eyes. Most females have a skewed X-inactivation pattern, with active selection, against the mutated X chromosome occurring (Kenwrick *et al.*, 2001).

10.8 Seeking consent from vulnerable individuals

In the example of DMD discussed above, determination of Sarah's carrier status was dependent on analyzing a sample from her brother. In syndromes in which mental retardation forms part of the phenotype, this may raise specific issues relating to consent.

Understanding the principles of consent and confidentially is central to ethical medical practice and forms part of many professional codes of ethics.

CASE STUDY – JIM

Jim provides a family tree that shows he has two sisters and one brother, Simon. The genetic counselor asks him about his sisters. Michelle is at university and his other sister, Kate, is still living at home. Kate had problems at school and needed some remedial help. She passed her exams and is working. She has a boyfriend and is planning to get married next year. Jim's mother had a brother who is in residential care with learning difficulties and the family has always thought that he and Jim's brother Simon looked identical. The reason for the learning difficulties was not known and the family had stopped looking for a diagnosis some time ago. The counselor explained that in order to advise Jim it would be helpful to have a diagnosis of Simon. This would mean him being seen and investigated, which might involve a blood test. As Simon was an adult this would have to be discussed with him.

Consent is required before undertaking any medical intervention and, to be valid, the consent has to be informed and obtained without coercion.

> 'In all cases, the *a priori* free and informed consent of the person concerned shall be obtained. If the latter is not in a position to consent, consent or authorization shall be obtained in the manner prescribed by law, guided by the person's best interest'.
>
> Article 5b of the Universal Declaration
> on the Human Genome and Human Rights

There are special issues when considering investigating a person who may be unable to consent either because of age, intellectual capacity or because there are legal constraints (such as for prisoners). The identification of who is able to give consent on behalf of another individual will be determined by local legal frameworks. For example, in England and Wales, at the time of writing, a person may not give legal consent on another adult person's behalf. If an individual is presumed unable to give informed consent because of mental incapacity, it is normal to assume consent if it can be shown that the procedure or treatment would be in the best interests of the person concerned.

Interpretation of this may be straightforward in terms of a medical investigation that is being considered to make a diagnosis that will affect the treatment or management of the person. It may be more difficult to make a decision in the case of a test that is being done for the benefit of someone else. In the case of individuals who have intellectual disabilities, it is also important not to assume that they are unable to understand enough to be able to consent to an investigation for the benefit of their relative. The practitioner may be able to present the information in such a way that the individual is able to understand and is able to give free and informed consent.

The issue of testing in children is one that has received some attention within the genetics counseling context and this is discussed separately in Chapter 4. It is important that the practitioner is aware of the legal, ethical, and professional frameworks that pertain to issues of consent to medical treatment as they apply in their area of practice.

10.9 Dynamic mutations in fragile X syndrome

As discussed, fragile X syndrome is caused by mutations in the FMR1 gene, which is located at Xq27.3, the fragile site identified in cytogenetic analysis. Nearly all the mutations are an expansion of a triplet repeat (CGG), resulting in abnormal methylation of the gene and failure to produce the protein. This repeat is in the 5′ untranslated region of the gene and is polymorphic in the general population, with a range of 6–60 repeats and a mode of 30. The abnormal methylation results in decreased or completely absent FMR1 protein. A minority of patients have deletions or point mutations.

Before the identification of the gene, it had been noted that there were some unusual features of the pattern of inheritance in families. For example, the gene seemed occasionally to be transmitted through unaffected grandfathers – males and, sometimes, females were affected. Affected females were, in general, more mildly affected (Sherman *et al.*, 1985).

The explanation was that the triplet repeat expansion – that is the causative mutation in fragile-x syndrome – was the first example of what is known as a 'dynamic mutation.' It changes in its transmission from generation to generation (Fu *et al.*, 1991). Analysis of pedigrees identified that, apart from the normal range of triplet repeats with no phenotypic effects, there are two other classes: a premutation allele that has a range of 60–200 repeats; and a full-mutation allele that has a range of more than 200 repeats. Premutation alleles are unstable and tend to expand when transmitted. The premutation can undergo a small expansion, usually resulting in a slightly larger triplet

Figure 10.6 Family tree showing transmission of premutations and full mutations

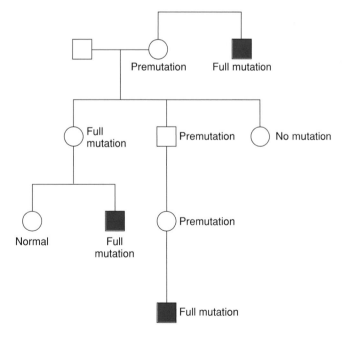

repeat, but still within the premutation range, or it can undergo a large expansion, resulting in a full mutation. This large expansion only seems to happen when a carrier female transmits the allele. The risk of expansion appears to relate to the size of the premutation, with the larger premutations being more unstable.

Thus these anomalous observations can be explained by the occurrence of premutations and full mutations within the family (Figure 10.6). A male can transmit the premutation to his daughters, who will be at risk of the allele expanding to a full mutation when they pass it on to their children.

CASE STUDY – JIM

After careful discussion with Simon, who indicates that he is aware he is different and wants to know why and is also willing to help his family, a blood sample is taken. The sample confirms that Simon has fragile X syndrome and the family and Simon are told the diagnosis. Jim is aware that there is no risk of him passing on a full mutation to his children, but does want to be tested for the premutation. The counselor organizes the appropriate testing and counseling for them all and discusses with them how they are going to tell the other family members who may be at risk of having children with the condition.

Molecular genetic testing in fragile X syndrome

The methods for testing for mutations in the *FMR1* gene will vary between different laboratories as no single technique will identify all mutation types.

PCR specific for the repeat region of the gene has a high sensitivity for repeat sizes in the normal and lower premutation range (Figure 10.7). However, it will fail to amplify larger repeat sizes. If the PCR shows a normal or premutation allele in males or two different alleles in females, further testing may not be indicated. However, there are cases of mosaicism for the expansion and further testing may be warranted.

Southern blot analysis detects the presence of all repeat sizes and allows for an estimation of the repeat number. Tests for methylation status will detect the full mutation.

The testing for fragile X syndrome is therefore complex and will depend on the purpose of testing: diagnostic, carrier or for population screening purposes. (Population screening for premutation carriers or in newborn screening for diagnosis has been discussed but has not yet been adopted as policy the UK or the USA. There are concerns about the predictive value of

Figure 10.7 Fragile X PCR testing

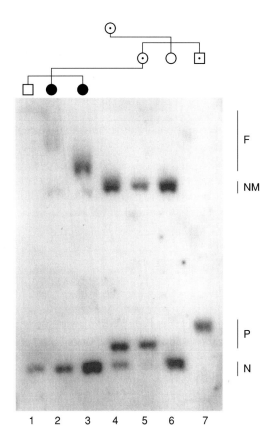

the test, particularly in females, the lack of a cure and the potential for causing harm.) The American College of Medical Genetics has issued a policy statement and technical guidelines for testing for fragile X syndrome in family members and in those in whom a diagnosis is suspected (American College of Medical Genetics, 1994; Maddalena *et al.*, 2001).

Special issues in fragile X syndrome

Mothers of boys with fragile X syndrome could be heterozygous carriers of a premutation or a full mutation. Full-mutation carriers can have physical and behavioral features that are similar to those in affected boys, but normally they occur at a lower frequency and are milder. This leads to special issues when counseling during prenatal diagnosis, when it is not possible to predict the phenotype of a girl with a full mutation. In addition, the methylation status of the *FMR1* gene may not be fully established by the time that a chorionic villus biopsy is performed and follow up amniocentesis may be required.

It has recently been established that carriers of premutations are at risk of other complications that are not seen in full-mutation carriers. Females are at risk of premature ovarian failure – that is, cessation of menstruation before the age of 40 (a recent review estimates the risk to be 20%: Murray, 2000).

Male carriers of the premutation (i.e. transmitting males) are at risk of a late-onset progressive cerebellar ataxia. The penetrance of this is age-related (Jacquemont *et al.*, 2003).

Diagnosis, testing and counseling in fragile X families is complex and requires the molecular laboratory clinician and counselor to liaise closely with each other and to be current in their knowledge base in order to provide accurate information for families.

10.10 Conclusion

In families with conditions that are X-linked, many of the issues relate to carrier testing in the female in the family. These questions may be answered using direct mutation analysis or by calculating individual risks. However, as we can see from the cases presented in this chapter, there are many complex issues emerging as more is discovered about the phenotypes associated with these mutations.

Study questions

Case scenario – Michelle

Michelle is referred from the obstetrician. She is 10 weeks pregnant and has a brother who has hemophilia. The obstetrican tells you on the phone that he has reassured Michelle because her clotting factors are within the normal range.

Task A
Is this reassuring? Write down what Michelle's risk might be.

Task B
Find out what testing might be available for Michelle and think about what her options might be.

Task C

Michelle's brother has a mild form of hemophilia, but sadly he is HIV-positive, having acquired the virus through contaminated Factor VIII.

What counseling issues may be important to consider?

References

American College of Medical Genetics (1994) Fragile X syndrome diagnostic and carrier testing. [Accessed July 28, 2004] http://www.acmg.net/resources/policies/pol-014.asp

Boland, B.J., Silbert, P.L., Groover, R.V., Wollan, P.C., Silverstein, M.D. (1996) Skeletal, cardiac, and smooth muscle failure in Duchenne muscular dystrophy. *Pediat. Neurol.* **14**: 7–12.

Covone, A.E., Lerone, M. and Romeo, G. (1991) Genotype-phenotype correlation and germline mosaicism in DMD/BMD patients with deletions of the dystrophin gene. *Hum. Genet.* **87**: 353–360.

Emery, A.E.H. (2002) The muscular dystrophies. *Lancet* **359**: 687–695.

Fu, Y.H., Kuhl, D.P., Pizzuti, A. *et al.* (1991) Variation of the CGG repeat at the fragile X site results in genetic instability: resolution of the Sherman paradox. *Cell* **67**: 1047–1058.

Grain, L., Cortina-Borja, M., Forfar, C., Hilton-Jones, D., Hopkin, J. and Burch, M. (2001) Cardiac abnormalities and skeletal muscle weakness in carriers of Duchenne and Becker muscular dystrophies and controls. *Neuromusc. Dis.* **11**: 186–191.

Heard, E., Lovell-Badge, R. and Avner, P. (1999) Anti-Xistentialism. *Nat. Genet.* **21**: 343–344.

Jacquemont, S., Hagerman, R.J., Leehey, M. *et al.* (2003) Fragile X premutation tremor/ataxia syndrome: molecular, clinical, and neuroimaging correlates. *Am. J. Hum. Genet.* **72**: 869–878.

Kenwrick, S., Woffendin, H., Jakins, T. *et al.* (2001) Survival of male patients with incontinentia pigmenti carrying a lethal mutation can be explained by somatic mosaicism or Klinefelter syndrome. *Am. J. Hum. Genet.* **69**: 1210–1217.

Maddalena, A., Richards, C.S., McGinnis, M.J. *et al.* (2001) Technical standards and guidelines for fragile X. *Genet. Med.* **3**: 200–205.

Murray, A. (2000) Premature ovarian failure and the FMR1 gene. *Semin. Reprod. Med.* **18**: 59–66.

Scriver, C.R., Tenenhouse, H.S. and Glorieux, F.H. (1991) X-linked hypophosphatemia: an appreciation of a classic paper and a survey of progress since 1958. *Medicine* **70**: 218–228.

Sherman, S.L., Jacobs, P.A., Morton, N.E. *et al.* (1985) Further segregation analysis of the fragile X syndrome with special reference to transmitting males. *Hum. Genet.* **69**: 289–299.

Turner, G., Webb, T., Wake, S. and Robinson, H. (1996) Prevalence of fragile X syndrome. *Am. J. Med. Genet.* **64:** 196–197.

Verkerk, A.J., Pieretti, M., Sutcliffe, J.S. *et al.* (1991) Identification of a gene (FMR-1) containing a CGG repeat coincident with a breakpoint cluster region exhibiting length variation in fragile X syndrome. *Cell* **65:** 905–914.

Further resources

American Medical Association statement on informed consent. [Accessed March 4, 2005] http://www.ama-assn.org/ama/pub/category/4608.html

Crawford, D.C. and Sherman, S.L. (2004) Fragile X syndrome: from gene identification to clinical diagnosis and population screening. In: *Human Genome Epidemiology.* (eds M.J. Khoury, J. Little, and W. Burke) Oxford University Press, New York, pp. 402–422.

Human Genetics Commission. Inside information 2002. Department of Health UK. (Chapter 4 of this report concentrates on consent and confidentiality as they relate to medical genetics.)

Kelley, L.A. (1999) *Raising a Child with Hemophilia – A Practical Guide for Parents.* Aventis Behring L.L.C. King of Prussia, PA, USA.

The National fragile X Foundation. [Accessed March 4, 2005] http://nfxf.org/

Familial cancer

11.1. What is cancer?

Cancer is a collective name for a number of diseases that are characterized by the growth of malignant cells. It is known to affect one in three people at some time during their lifetime (Cancer Research UK, 2004). As it is so common, particularly among the elderly, many individuals will have some family history of cancer. Although treatment of cancer has improved dramatically and survival rates for many types of cancer (such as breast, stomach, and colon cancer) are far better than they were a generation ago (Cancer Research UK, 2004), for many people a diagnosis of cancer is interpreted as a death sentence. It is understandable that a family history of cancer can be the source of severe anxiety in some clients.

Cancer is a term used to describe a group of diseases in which there is cell growth that is not regulated by the normal controlling mechanisms. Cancers are identified and named according to the type of tissue from which they originate, and any organ may therefore be affected by a variety of different types of cancer. Table 11.1 shows the types of cancer that may develop according to the tissues from which they originate. Cancer *in situ* is a term that is used to describe a tumor that has remained within the organ of origin, without spreading to other parts of the body. Spread of the malignant cells may occur locally into the surrounding connective tissue, from which cells may travel via blood, lymphatic fluid or other body fluids (such as cerebrospinal fluid). Secondary tumors are termed metastases. In confirming cancer diagnoses within a family, it is important to determine the site of the primary tumor, rather than the metastases.

Table 11.1 Classification of tumors according to tissue type

Type of tissue	Classification of tumor	Tumor example
Epithelium (e.g. skin, and lining and covering of organs such as bowel, lung, breast)	Carcinoma	Adenocarcinoma
Muscle	Myosarcoma	Rhabdomyosarcoma
Nervous tissue	Glioma or neuroblastoma	Optic glioma
Connective tissue (e.g. cartilage, bone, fat)	Sarcoma	Osteosarcoma
Endothelium (e.g. veins and arteries, meninges)		Angiosarcoma, Meningioma
Hemopoietic/lymphoid tissue (e.g. bone marrow)		Leukemia or lymphoma
Germ cells (e.g. ovary or testicle)		Teratoma

11.2 Inherited susceptibility to cancer

CASE STUDY – MARION

Marion Walker is a 38-year-old woman, divorced, with three children. She sees her family doctor regularly for a prescription for the oral contraceptive pill. At one visit, she is tearful and tells her doctor that her mother has been told she has secondary cancer and has only a few months to live.

Her doctor asks about the family history of cancer, and Marion tells her that her mother was diagnosed with breast cancer 4 years earlier and had a mastectomy and chemotherapy at that time. The family had believed the cancer had been beaten, and it had come as a shock when her mother was found to have bone metastases in her upper spine. The family was reeling, especially as a cousin had died only 2 months ago from breast cancer and another aunt had died suddenly from ovarian cancer several years earlier. Marion feels the family is blighted by cancer and 'she will be the next to go'.

Marion's doctor suggests referring her to the local genetics service for assessment. Although Marion is worried about herself, she finds it impossible to talk about the family and defers her appointment.

Marion is typical of many clients referred to a genetics service because of their worry about their family history of cancer. Although inherited predisposition to cancer has only been demonstrated at a molecular level in the past few decades, families have long been aware of increased susceptibility to cancer. Lynch and Lynch (1994) cite the case reported by Aldred Warthin of a seamstress concerned about her own risk of cancer in 1895 because of her family history. In this case, the belief of the seamstress that she would die young was justified; she died from endometrial cancer and may have carried an hereditary non-polyposis colon cancer (HNPCC) gene mutation. However, the perception exists in some members of the general population that cancer is a single disease entity, and that any cancer history may increase general susceptibility to malignancy in family members. It should also be emphasized that most cancers are not due to an inherited gene mutation. Environmental influences play a larger role. Deaths from lung cancer account for about one-third of all cancer deaths in the UK, and cigarette smoking is implicated in 90% of all deaths from lung cancer (Cancer Research UK, 2004).

11.3 The causes of cancer

Cancer arises as a result of a sequence of events at cellular level. The regulation of cell growth is accomplished by a complex set of processes, and genes control the initiation of cell division (proto-oncogenes) and the arrest of that division when sufficient new cells have been produced (tumor-suppressor genes). When DNA is being replicated, any damage to the sequence is usually repaired by a mismatch repair (MMR) gene. In cases of severe DNA damage, the cell will die due to the process of apoptosis, or programmed cell death. If the MMR gene itself is mutated, damaged DNA will not be repaired or destroyed, and the resultant DNA sequences may replicate themselves into the new cell, with high potential for malignant change. The p53 tumor-suppressor gene is a key MMR gene, and mutations in this gene are found in many malignant cells (Lynch *et al*, 1998). Exposure to carcinogens, such as cigarette tar, radiation, ultraviolet light, and some viruses, has been shown to contribute to the changes that eventually result in malignancy (Alexander *et al.*, 2000).

Knudson's hypothesis (two-hit theory)

Knudson (1971) proposed a theory about the development of cancer, based on observations about the condition retinoblastoma. The theory proposes that cancer develops after mutations occur in both copies of a particular gene (Figure 11.1). The gene involved in retinoblastoma is called the *Rb* gene, and

Figure 11.1 **Two-hit theory**

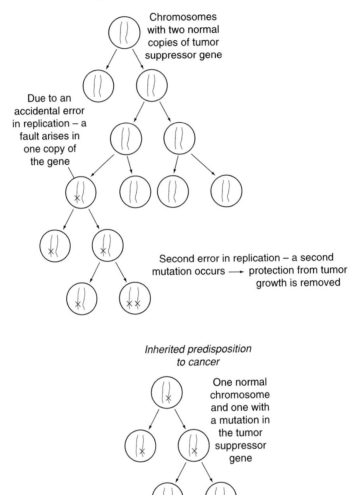

Chromosomes with two normal copies of tumor suppressor gene

Due to an accidental error in replication – a fault arises in one copy of the gene

Second error in replication – a second mutation occurs ⟶ protection from tumor growth is removed

Inherited predisposition to cancer

One normal chromosome and one with a mutation in the tumor suppressor gene

Second gene mutation occurs ⟶ Protection from tumor growth removed

the condition is usually seen in children, even neonates. In some children, two random and sporadic changes in the *Rb* gene cause the loss of function of the gene, but if a child inherits a single mutated copy of the gene, only one other sporadic mutation event has to occur for the tumor to develop. A child who inherits a mutation from a parent may develop tumors in both eyes due to a highly increased susceptibility.

This theory helps to explain why individuals who have inherited a gene mutation develop a malignancy at a lower mean age than those in whom the development of cancer relies on two sporadic mutations. Mutations are more likely to occur as the number of cell divisions increases throughout a person's lifetime, so the older the individual, the greater the number of cell divisions and the more likely a mutation is to occur.

11.4 Genetic counseling for familial cancer

Genetic assessment for familial cancer

The task of the health professional working with clients who have a family history of cancer is to discern those families who have an inherited predisposition from those whose cancer has occurred sporadically, albeit in several family members.

The purpose of this assessment is to:

- inform family members of their risk as accurately as possible;
- offer screening to detect malignancy when indicated;
- enable those at risk to consider preventive treatment (such as prophylactic surgery), if appropriate.

In this chapter, familial cancer will be discussed according to the level of risk of an individual in the family to develop cancer. This is dependent on whether there is evidence in the family of:

1. A single gene disorder that is likely to manifest in malignancy.
2. An inherited gene mutation that increases susceptibility to cancer.

11.5 Single gene disorders

There are a number of genetic conditions caused by mutations in a single gene that cause malignancy in a person who carries the mutation. These conditions may follow a pattern of autosomal-dominant (see Chapter 8) or autosomal recessive inheritance (see Chapter 9). The condition von Hippel Lindau disease is a dominantly inherited condition and is discussed in detail here. It may be the responsibility of a genetics service to organize appropriate screening programs for patients who are affected by or at risk of these multi-system conditions.

von Hippel Lindau disease

Von Hippel Lindau (VHL) disease is caused by a mutation in a tumor-suppressor gene, called the *VHL* gene. The protein product of this gene, pVHL, regulates a transcription factor (hypoxia-inducible factor, HIF) that has a role in the control of vascular endothelial growth factor (VEGF) production (George and Kaelin, 2003). Individuals affected by VHL may develop vascular tumors, such as retinal hemangiomas, cerebellar hemangiomas, and phaechromacytomas, as well as cysts in the kidney and liver, and renal-cell carcinoma. Genetic testing is available, but the gene mutation cannot be identified in all familial cases. If an individual has inherited the mutation, or is known to be at risk, then a program of screening should be implemented to detect potential complications of the disease and so enable treatment to be prescribed. The condition may manifest itself from childhood, and screening needs to be introduced from infancy.

The screening protocol varies from center to center, but broad guidelines are given here. However, the frequency and type of screening will alter if abnormalities are detected in any individual. As the condition affects so many body systems (Table 11.2), it may be a function of the genetics service to organize or coordinate the screening program. A genetics registry is

Table 11.2 Clinical surveillance in VHL syndrome

Clinical risk	Screening test	Frequency	Commencing at age
Renal cysts Renal-cell carcinoma Hepatic cysts	Abdominal scan (ultrasound or CT)	Ultrasound: annually CT: every 1–3 years	11–14 years
Retinal hemangioma	Slit lamp eye examination by ophthalmologist	Annually	From 1–5 years
Cerebellar hemangioma	Neurological examination Enhanced brain scan (MRI)	Every 2–3 years until 50–60 years, then every 5 years.	11–14 years
Phaechromacytoma	24-hour urine collection to measure catecholine levels	Annually	2 years

frequently used to ensure that affected or at-risk individuals are recalled for screening and genetic counseling at appropriate times, and a joint clinic involving all of the relevant medical specialists may reduce the need for families to attend for screening on multiple occasions within the year.

Treatment options   Retinal hemangiomas can be treated with laser therapy but may result in loss of vision. Renal and hepatic cysts may not require specific treatment. There are surgical difficulties in removing cerebral hemangiomas and therefore these may be treated conservatively unless they reach a certain size, when surgery may become necessary. Renal cell carcinoma is usually treated aggressively with surgery.

Genetic risks   As VHL is an autosomal dominant condition, each child of an affected person is born at a 50% risk of having the condition and should be offered screening unless mutation analysis demonstrates that the person has not inherited the mutation that is present in affected members of his or her family. There is some evidence that when the condition is diagnosed in the family for the first time, mosaicism may be present in the affected person (Sgambati et al., 2000). In these cases, siblings may need to be offered testing even if no mutation is detected in either of the parents of the affected individual.

Neurofibromatosis type I

A mutation in the neurofibromatosis type I (NF1) gene (17q11.2) can produce a range of clinical signs and symptoms in those affected by NF1 (Szudek et al., 2003). In many patients it is a benign condition, causing only skin manifestations, such as café au lait patches, that appear in early childhood. Neurofibromata are lumps on the skin, which are caused by overgrowth of the neural coverings; these may start to appear in adolescence. They are benign but may cause serious problems if located in a position where space for growth is restricted, such as around a spinal nerve. Nerve sheath tumors, such as a plexiform neurofibroma (PNF), may become malignant but survival is enhanced if the tumor is detected early and it is therefore advised that any PNF should be monitored using MRI scans (Mautner et al., 2003). Some patients develop other types of malignancy of nervous or connective tissue, such as glioma or sarcoma.

Other potential effects of being heterozygous for the NF1 gene are macrocephaly, learning difficulties, scoliosis, and kyphosis. Affected patients may have Lisch nodules in the eyes; these are harmless but can be a diagnostic aid.

A diagnosis is usually made on the basis of clinical examination and family history, although many cases are new mutations within the family.

The role of the genetics nurse or counselor may involve organization of screening as appropriate. However, the client may be distressed by the cosmetic appearance of the neurofibromata, particularly if these are prominent around the face, and psychological support should be offered either through the genetics team or referral elsewhere.

Neurofibromatosis type II

Neurofibromatosis type II (NF2) is a completely different condition to NF1. The tumor-suppressor gene involved has been located at 22q12.2. Affected persons may suffer serious tumors of the nervous and endothelial tissue, such as vestibular schwannoma and meningioma.

Ataxia telangectasia

Ataxia telangectasia (AT) is a rare, recessively inherited condition. The karyotype of an affected person shows the characteristic broken chromosome appearance. The cancers develop because of the disruption to the MMR function within the DNA molecule, rather than because of a single mutation in an oncogene or tumor-suppressor gene, and therefore cancers can occur in many organs. For this reason, it is unfortunately not practical to screen such patients for gene mutations.

Li Fraumeni syndrome

This dominantly inherited condition is due to the inheritance of a mutation in the *p53* gene (17p.13.1). This type of mutation should be suspected when the family reports a strong history of unusual tumors occurring at relatively young ages. These could include tumors of the brain and central nervous system (e.g. astrocytoma, medullablastoma, adrenocortex), breast, connective tissue (e.g. osteosarcoma), or hemopoetic system.

Familial adenomatous polyposis

Familial adenomatous polyposis (FAP) is a dominantly inherited condition that is caused by a fault in the *APC* gene on chromosome 5q21-q22. The mutation influences the development of multiple adenomatous polyps on the colon wall. The polyps are initially benign but malignant changes occur at the cellular level, and small adenomas then develop into full-scale carcinoma if untreated.

A small number of colonic polyps is found in a percentage of the population, but individuals affected by FAP develop hundreds of polyps (or more). A count of 100 or more polyps is diagnostic of FAP. Polyps may begin to develop in early childhood, but frequently the onset of polyps occurs during early adolescence. For this reason, screening is advised from around 10–12 years of age on an annual basis. Colonoscopy, a test that enables the colon to be directly viewed through an endoscope, is considered to be the test of choice. If polyps are seen at colonoscopy, they can be removed at the time, usually using laser treatment. Polyps may also occur in the small intestine and therefore upper gastro-intestinal tract screening is also advised for these patients.

As the number of polyps in the colon increases during adolescence, it becomes impossible to remove each one individually and a colectomy to remove the colon is required. This is best performed as a planned procedure, when the patient is well and psychologically prepared for the removal of the colon. Some surgeons advocate early surgery during childhood, to enable the child to adjust both mentally and physically, but others feel it is important to enable the individual to understand the surgery and give informed consent.

As a result of the colectomy, the affected person may have a colostomy, especially if polyps form in the rectum. However, increasingly an ileo-rectal pouch procedure is performed to avoid the need for a colostomy. This involves using a part of the small intestine to form a new rectal cavity. A temporary colostomy may be required at the time of initial surgery to facilitate healing.

The presence of polyps in the colon of an individual with a family history of FAP is diagnostic of the condition. However, gene mutations can be found in the majority of families, and predictive genetic testing avoids the need for colonoscopic screening in at-risk individuals. As screening usually starts in early adolescence, genetic testing may be offered at the age of 10–12 years. Whereas this is younger than the age at which predictive testing would normally be offered – as the result would influence medical management of the child – it is generally considered to be good practice. However, every effort should be made to ensure that the child understands the test and its implications and that the child's consent is obtained as well as parental consent.

11.6 Inherited gene mutations that increase susceptibility to cancer

General information

Individuals who inherit a mutation in either a tumor-suppressor gene or an oncogene may have an increased predisposition to development of malignancy in many organs. However, the inherited mutation simply increases the predisposition to cancer, and development of malignancy depends on a second mutation occurring, in addition to other events at the cellular level. A person who has a mutation may therefore never develop a cancer. It is important to remember this when taking a family history, as the intermediate unaffected relative (between two generations of individuals who have been affected with cancer) may still have the mutation.

Familial breast/ovarian cancer

Familial breast cancer is known to be influenced by mutations in tumor-suppressor genes. When functioning normally, these genes have a protective function against cancer. Mutations are detectable in some families in the *BRCA1* and *BRCA2* genes. Researchers continue to search for other high-penetrance genes that increase susceptibility to breast cancer (Thompson *et al.*, 2002). In addition, many gene variants in low-susceptibility genes may moderately increase the risk of breast or ovarian cancer (Wooster and Weber, 2003).

The risks of breast cancer conferred by these mutations must be viewed against the lifetime population risk of breast cancer, which is approximately 1 in 9 (10.9%) women in the UK population (Cancer Research UK, 2004) and approximately 1 in 8 (13.3%) women in the USA (National Cancer Institute, 2004). Genetic healthcare practitioners need to be aware of the baseline level of risk for their own population.

BRCA1   The BRCA1 gene is located at chromosome locus 17q21 (OMIM, 2004). Carriers have a 74% lifetime risk of developing breast cancer, and a lower risk of developing ovarian cancer (48%), although the risk does seem to vary with the type of mutation (Pharoah, 2002).

BRCA2   In a number of studies, carriers of the BRCA2 mutation have been shown to have a risk of developing breast cancer during their lifetime of 69%, and a risk of developing ovarian cancer of 12% (Antoniou *et al.*, 2003). Men who inherit the mutation have an increased risk of developing breast cancer and are slightly more prone than other men to develop prostate cancer. There is thought to be an association between *BRCA2* and an increased risk of pancreatic cancer.

In many cases, a familial tendency is observed, but the gene mutation cannot be identified in the family. This is because the condition is heterogeneous and there are many possible mutations in each potential gene. It is preferable to focus initially on a sample from an affected family member, to maximize the chance of finding a mutation. Families in whom a mutation cannot be identified, cannot be reassured that a mutation does not exist. Where there is a strong family history, screening should be offered even when a mutation is not found in an affected family member (Eccles *et al*, 2000).

Although, in the past, the psychological preparation and support of affected family members was not a priority, affected women may need support to adjust to the result. Women who have experienced breast cancer are not always prepared to be told that they had a mutation, a result that has implications for their own future health as well as the health of their family members. Some carry a feeling of guilt at having passed on the mutation.

It is generally considered good practice to offer counseling to a client who requests pre-symptomatic genetic testing in order to enable them to explore the implications of the test outcome and to prepare psychologically for the result. The nurse or counselor should facilitate the client to consider:

- the reasons for requesting the test;
- the reaction to either a positive or negative test result;
- plans for screening or prophylactic surgery if the result is positive;
- the implications for financial status and insurance;
- the implications of the result for relatives and other close contacts;
- psychological coping strategies;
- support mechanisms;
- sharing the news – who will they tell?

Post-test follow-up counseling should also be offered regardless of the test outcome. It has been demonstrated that individuals who have a negative result, and are aware of having escaped the familial condition, may have difficulties adjusting and might experience survivor guilt. Women who have a positive result may wish to consider strategies for prevention or screening in the context of having their increased risk confirmed.

Screening for women at increased risk of breast cancer   Following the publication of guidelines by the National Institute for Clinical Excellence (NICE, 2004; see Table 11.3), the following screening protocol is considered appropriate for women in the UK. However, screening programs differ enormously across countries and even in different regions of countries. Some women who are eligible for screening will choose not to be screened as they find the process itself psychologically traumatic and prefer to avoid thinking about their risk if possible.

Guidelines and software packages, such as Cyrilic, are available to support the practitioner to assess the level of risk for a women with a family history of cancer. As a general guideline, a woman would be considered to be in the moderate risk group (lifetime risk of breast cancer of 17–30%) if she has any one of the following:

- one first-degree relative affected with breast cancer at less than 40 years old;
- two relatives affected, at least one being a first-degree relative;
- at least three relatives affected with breast cancer;
- a male relative with breast cancer;
- a relative with bilateral breast cancer.

Women with more affected relatives may be in the high-risk group (more than a 30% lifetime risk of breast cancer). Note that the presence of relatives with bilateral breast cancer, breast and ovarian cancer, or male breast cancer generally increases the risk of a breast cancer gene mutation being present in the family (NICE, 2004).

Table 11.3 Current UK recommendations for screening for breast cancer risk (NICE, 2004)

Group	Recommended screening
Women over the age of 50 years at low risk of breast cancer	Mammogram every 3 years from the age of 50 years
Women of age 40–49 years and at moderate or high risk of breast cancer	Yearly mammogram
Women of age 30–39 years and at increased risk of breast cancer	Mammogram only if part of individualized screening strategy

Screening/prophylaxis for women at increased risk of ovarian cancer

Women who have a family history that is suggestive of a *BRCA* mutation, or who have been identified as having a *BRCA1* or *BRCA2* mutation, should be offered screening for ovarian cancer. If the woman has no plans to have future pregnancies, she may consider bilateral prophylactic salpingo-oopherectomy as an option to reduce the risk of both breast and ovarian cancer (Rebbeck, 2002). However, the surgical menopause that this surgery induces can have adverse health effects (i.e. increased risk of osteoporosis) if performed at a young age, and extended use of hormone replacement therapy (HRT) has been shown to increase the chance of the woman developing breast cancer. However, the risk of breast cancer in a woman who has had salpingo-oopherectomy and HRT therapy is still lower than the risk to the same women without these interventions (Rebbeck *et al.*, 1999).

Current screening for ovarian cancer has not yet been proven to be effective, but women may be offered an ultrasound scan of the ovaries (via the vagina) and testing of CA125 levels. Elevated CA125 levels may be due to reasons other than ovarian cancer. Ovarian biopsy is performed if there are abnormalities detected in either screening test. However, ovarian cancer may not be detected using these methods.

Advice for men who have a *BRCA1* or *BRCA2* mutation

Men who have a *BRCA* mutation or strong family history of breast cancer have less chance of developing breast cancer than their female relatives. However, they have an increased risk of developing breast cancer compared with other males and should be advised to be aware of any lumps, thickening, or other changes on the chest wall or armpit, and to seek medical advice if they notice changes. As their risk of developing prostate cancer is higher than for men in the general population, annual prostate-specific antigen (PSA) screening and digital rectal examination of the prostate may be offered; however, the sensitivity of PSA testing in detecting prostate cancer is still not proven and there is still some controversy as to whether this is a helpful screening test (NSC, 2004).

Excellent updated information on familial breast cancer can be found on the Gene Reviews and National Cancer Institutes websites (see 'Web-based resources').

CASE STUDY – MARION

Six months after the original referral to the genetics service, the genetic counselor writes to Marion again, asking if she would like an appointment. Marion now feels ready to discuss the issues. The counselor sends her a letter explaining the service and asks her to provide written information about the family history. Written consent from Marion's mother is required before her medical notes can be viewed. No consent is required for access to the notes of deceased relatives. Marion is sent an appointment to see the genetic counselor after the medical history and cancer diagnoses in her relatives has been confirmed.

The relevant details are:

Marion's mother, Helen, was diagnosed with ductal carcinoma of the breast at the age of 55 years. She is confirmed to have bone metastases in the spine.

Marion's maternal aunt, Faye, had metastatic ovarian cancer at the age of 57 years. The primary tumor could not be identified during her illness but was found at post-mortem examination.

Faye's daughter, Melanie (Marion's cousin), had been diagnosed with ductal carcinoma of the left breast at age 34 years. She had died at 39 years after 5 years of treatment.

Family tree

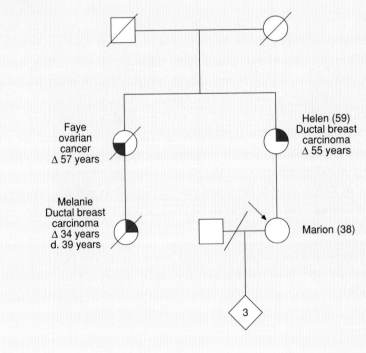

Marion's family history is highly suggestive of a *BRCA* gene mutation. She is advised that she should commence both breast and ovarian screening, and is referred to the Breast Service and the gynecologist who runs the ovarian cancer screening service.

The subject of predictive testing for Marion is raised. At present, she does not wish to be tested but feels she might if she has to make a decision about prophylactic oopherectomy.

Marion asks her mother if she would provide a sample of blood for DNA studies. Helen is keen to help her daughter in any way she can, but does not want to know the result. She is aware she is dying, and agrees that the sample can be used after her death for genetic mutation studies.

Sadly, Helen dies 3 months later.

One year later, Marion asks the genetic counselor to send her mother's sample to the laboratory for analysis. Within 3 months, a *BRCA2* mutation has been found in her mother's sample.

Marion delays testing for herself for another year, but then wants to decide about having an oopherectomy. She is concerned that screening for ovarian cancer is not always effective and she does not want to leave her teenage children without a mother. Marion is tested and does not have the mutation.

Colorectal cancer

The dominant condition of FAP has been discussed in section 5. This is an important genetic cause of colorectal cancer, but is generally classified as a single gene disorder.

Hereditary non-polyposis colon cancer This condition is slightly misnamed, as clients with HNPCC sometimes have small numbers of colonic polyps and may have cancer in organs other than the colon. About 1–5% of colorectal cancers are due to a germline HNPCC mutation (Moslein, 2003). HNPCC is also sometimes referred to as Lynch syndrome. Tumors of the colon, rectum, endometrium, ovary, stomach, and kidney have been associated with HNPCC mutations.

A number of MMR genes are implicated in HNPCC. The most common are *hMSH2* (on 2p15) and *hMLH1* (on 3p21), but there are at least three others that have been identified in families affected by HNPCC.

Microsatellite instability (MSI) of the DNA is frequently observed in tumors that are associated with a HNPCC mutation (Chung and Rustgi, 2003), and

Box 11.1 Amsterdam criteria for HNPCC

- Histologically confirmed colorectal cancer in at least three relatives, one of whom is a first-degree relative of the other two.
- Occurrence of disease in at least two successive generations.
- Age at diagnosis below 50 years in at least one colorectal cancer case.
- Exclusion of FAP.

testing the tumor for MSI can be used as a guideline for further mutation analysis.

To identify families who were likely to have a HNPCC mutation, and therefore to require cancer screening, the Amsterdam criteria were drawn up (Box 11.1).

However, it was felt that these criteria were not sufficiently sensitive to detect all potential HNPCC families, and the Bethesda criteria are now used by many centers (Box 11.2).

If there is sufficient evidence to suspect the presence of an HNPCC mutation in the family, then a DNA sample from an affected family member may be analyzed to detect a mutation. In such a heterogeneous condition, with many potential mutations, this is often a lengthy process. Testing a tumor sample for MSI can provide guidance as to the likelihood of a mutation being present.

Box 11.2 Bethesda criteria for HNPCC

- Individuals with a colorectal cancer family history that meets the Amsterdam criteria.
- Individuals with colorectal cancer at 45 years old or younger, or endometrial cancer at 45 years old or younger, or a colorectal adenoma at 40 years old or younger.
- Individuals with colorectal cancer and a first-degree relative with colorectal cancer or an HNPCC-related cancer; one of the cancers diagnosed at 45 years old or younger. (HNPCC-related cancers include colorectal, signet-ring-cell-type colorectal, endometrial, stomach, biliary tract, urinary tract, ovarian, and skin cancers.)
- Individuals with two HNPCC-related cancers, including synchronous and metachronous colorectal cancer.

Box 11.3 HNPCC screening guidelines

- Colorectal (all cases) colonoscopy every 2 years at age 25–75 years.
- Endometrial* annual pipelle biopsy (suction curettage) and ultrasound at age 30–65 years (not yet proven to be of benefit).
- Ovarian# annual transvaginal ultrasound and serum CA125 concentration at age 30–65 years.
- Transitional-cell carcinoma in the urinary tract* annual hematuria test at age 25–40 years; annual urine cytology at age 40–65 years (with or without cystoscopy every 1–2 years); annual renal ultrasound at age 40–65 years.

*Families with history of endometrial cancer and mutation-positive families.
#Families with history of ovarian cancer.
(Cole and Sleightholme, 2000).

In an individual who is thought to be at high risk or proven to have an HNPCC mutation, various screening guidelines are applicable (Box 11.3).

Family history of colorectal cancer not fulfilling criteria for HNPCC

In many cases, the family history will indicate an increased risk of colorectal cancer, but will not fulfill the HNPCC criteria. In these cases, screening may be recommended based on the level of risk.

Lifetime risk assessments can be made, based on the number and relationship of affected relatives (Box 11.4).

Recommendations for colonoscopic screening vary from country to country. In the UK, the Cancer Family Study Group (Hodgson *et al.*, 1995) recommended colonoscopy every 4–5 years from the age of 25 years old (increasing to every 3 years if a malignancy is detected) for individuals with

Box 11.4 Lifetime risk of colorectal cancer

Population risk	1 in 30
One first-degree relative affected (any age)	1 in 17
One first-degree and one second-degree relative affected	1 in 12
One first-degree relative affected (age < 45 years)	1 in 10
Two first-degree relatives affected	1 in 6
Autosomal-dominant pedigree	1 in 2

(Houlston *et al.*, 1990)

an estimated lifetime risk greater than 10%. Individuals whose family history suggests a lower risk should not undergo colonoscopic screening, but a lower-risk procedure, such as faecal occult blood testing, may be justified. In the USA, universal population screening using colonoscopy has been recommended (Winawar *et al.*, 2003) from the age of 50 years.

11. 7 Clinical practice

The family cancer clinic

The assessment of genetic risk relies on the accuracy of reported cases of cancer in the family. The reporting of breast cancer cases has been shown to be accurate in the majority of cases (Husson and Herrinton, 2000), but in all other cancers it is wise to document the type and site of the cancer and the age at which the affected person was diagnosed.

This can be done via the cancer registries, medical records or by means of the death certificate. However, due to concerns about confidentiality and privacy, written consent is required from living relatives before their records can be accessed.

After taking an accurate family tree and documenting all significant medical history, a risk assessment of the likelihood of the client developing cancer can be made. Recommendations for screening can be made based on the level of risk and the local screening arrangements. Additional health promotion information about diet, exercise, smoking cessation, normal weight control, use of the oral contraceptive pill, and hormone replacement therapy can also be offered to the family. Genetic testing is discussed if available and appropriate. Appropriate referrals are made for screening and/or discussion of prophylactic surgery. The family is asked if they are able to discuss risks with other family members who may be at risk. Support may be needed from the genetics team to provide information and offer contact. The family are asked to return to the genetics service again if they become aware of any new information that might alter the risks, so a revised assessment can be made. A summary letter including all relevant information is provided for the family, the family doctor and the referrer.

11.8 Psychological aspects of familial cancer risk

The client with a strong family history of cancer has frequently had to deal with a number of family deaths due to the condition. Mourning for each illness or death may not be completed before the next one occurs and this can result in complicated grief (Worden, 1991), which is covered in Chapter 5.

In addition to the burden of grief, the client may have particular lay beliefs about his or her own predisposition for cancer (Skirton and Eiser, 2003). For example, a woman who looks very much like her affected mother may believe her risk is higher than that of her sister. Anxiety is often particularly high around the time the client reaches the age at which a close relative was diagnosed. Families may have 'preselected' those they believe are affected (Kessler, 1998), making it difficult to adjust if risk estimates or genetic test results contradict those beliefs. This situation requires skilful counseling to support the family in integrating the new information into their previously held beliefs (Skirton, 2001). Those whose test results are negative may experience survivor guilt at having escaped the family 'curse'.

Anxiety about cancer risk may produce changes in self-monitoring behavior. In some clients, anxiety produces a tendency to over-monitor; for example, checking breasts for lumps every day. Other clients report a reduction in monitoring after becoming aware of their high risk because the fear of actually finding an abnormality is so great (Lindberg and Wellisch, 2001). Assisting clients to manage the psychological effects of having a family history of cancer are part of the role of the genetic healthcare practitioner.

11.9 Conclusion

Cancer is a common condition, and only a minority of cancers are due to an inherited genetic mutation. However, accurate risk assessment and mutation testing can provide the information that is needed by a family to promote health, reduce the risk of cancer and maximize the chance of survival through early detection of tumors.

Study questions

1. Case scenarios

Based on the information given, are the following clients at significant risk of an inherited cancer predisposition?

(a) Lynne's mother died from bilateral breast cancer at age 38 years but she does not have other relatives who are affected with cancer.

(b) Brian's father had cancer of the stomach at the age of 67 years; his mother had cancer of the colon at the age of 41 years.

(c) Helen had a brother with cancer of the pancreas at the age of 56 years, a sister with lung cancer at the age of 48 years (a heavy smoker) and a mother with endometrial cancer at the age of 60 years.

(d) Mohammed's father had colon cancer at the age of 53 years; his sister had endometrial cancer at the age of 47 years and his uncle (father's brother) had colon cancer at the age of 49 years.

(e) Barry's sister had breast cancer at the age of 39 years; his mother had ovarian cancer at the age of 56 years and his maternal grandfather had pancreatic cancer at the age of 59 years.

2. Case scenario – James

James is 28 years old. He requests a genetic referral to discuss his worries about colon cancer. His father died from colon cancer at the age of 37 years; his mother is still living. James thinks his paternal aunt died from cancer of her 'female parts.' His father's parents both died in their 50s of cancer.

Task A
List the essential information you need about James' family history to make a genetic risk analysis for him.

Task B
James has no living affected relatives. What are his options with regards to genetic testing?

Task C
After your enquiries, you assess James' lifetime risk of having inherited a mutation that predisposes him to colorectal cancer as 1 in 2. What are the recommendations for colorectal cancer screening in your own region for someone like James?

References

Alexander, M.F., Fawcett, J.N. and Runciman, P.J. (2000) *Nursing Practice Hospital and Home*, 2nd Edn. Churchill Livingstone, London.

Antoniou, A., Pharoah, P.D., Narod, S. *et al.* (2003) Average risks of breast and ovarian cancer associated with BRCA1 or BRCA2 mutations detected in case series unselected for family history: a combined analysis of 22 studies. *Am. J. Hum. Genet.* **72**: 1117–1130.

Cancer Research UK (2004) What is cancer? [Accessed March 4, 2005] http://www.cancerresearchuk.org/aboutcancer/

Cole, T.R. and Sleightholme, H.V. (2000) ABC of colorectal cancer. The role of clinical genetics in management. *BMJ* **321:** 943–946.

Chung, D.C. and Rustgi, A.K. (2003) The hereditary nonpolyposis colorectal cancer syndrome: genetics and clinical implications. *Ann. Intern. Med.* **138:** 560–570.

Eccles, D.M., Evans, D.G.R. and Mackay, J. on behalf of the UK Cancer Family Study Group (UKCFSG) (2000) Guidelines for a genetic risk based approach to advising women with a family history of breast cancer. *J. Med. Genet.* **37:** 203–209.

George, D.J. and Kaelin, W.G. (2003) The von Hippel–Lindau protein, vascular endothelial growth factor and kidney cancer. *N. Engl. J. Med.* **349:** 419–421.

Hodgson, S.V., Bishop, D.T., Dunlop, M.G., Evans, D.G.R. and Northover, J.M.A. (1995) Guidelines developed by the UK Cancer Family Study Group. *J. Med. Screening* **2:** 45–51.

Houlston, R.S., Murday, V., Harocopos, C., Williams, C.B. and Slack, J. (1990) Screening and genetic counseling for relatives of patients with colorectal cancer in a family cancer clinic. *BMJ* **301:** 366–368.

Husson, G. and Herrinton, L.J. (2000) How accurately does the medical record capture maternal history of cancer? *Cancer Epidemiol. Biomarkers Prev.* **9:** 765–768.

Kessler, S. (1998) Invited essay on the psychological aspects of genetic counselling. V. Preselection: a family coping strategy in Huntington's Disease. *Am. J. Med. Genet.* **31:** 617–621.

Knudson, A.G.Jr. (1971) Mutation and cancer: statistical study of retinoblastoma. *Proc. Natl Acad. Sci. USA* **68:** 820–823.

Lindberg, N.M. and Wellisch, D. (2001) Anxiety and compliance among women at high risk for breast cancer. *Ann. Behav. Med.* **23:** 298–303.

Lynch, J. and Lynch, H.T. (1994) Genetic counseling and HNPCC. *Anticancer Res.* **14:** 1651–1656.

Lynch, H.T., Casey, M.J., Lynch, J., White, T.E.K. and Godwin, A.K. (1998) Genetics and ovarian cancer. *Semin. Oncol.* **25:** 265–280.

Mautner, V.F., Friedrich, R.E., Von Deimling, A., Hagel, C., Korf, B., Knofel, M.T., Wenzel, R. and Funsterer, C. (2003) Malignant peripheral nerve sheath tumours in neurofibromatosis type 1: MRI supports the diagnosis of malignant plexiform neurofibroma. *Neuroradiology* **45:** 618–625.

Moslein, G. (2003) Clinical implications of molecular diagnosis in hereditary nonpolyposis colorectal cancer. *Recent Results Cancer Res.* **162:** 73–78.

National Cancer Institute. (2004) Cancer Facts [Accessed March 4, 2005] http://www.nci.nih.gov/

NSC (2004) National Screening Committee guidelines. [Accessed March 4, 2005] http://www.nsc.nhs.uk/

NICE (2004) Familial breast cancer. The classification and care of women at risk of familial breast cancer in primary, secondary and tertiary care. [Accessed March 4, 2005] http://www.nice.org.uk/page.aspx?o=203181

OMIM (2004) BRCA1. [Accessed March 4, 2005] http://www.ncbi.nlm.nih.gov/entrez/dispomim.cgi?id=113705

Pharoah, P.D. (2002) Penetrance of BRCA1 is high. Minerva. *BMJ* **325:** 502.

Rebbeck, T.R., Levin, A.M., Eisen, A. *et al.* (1999) Breast cancer risk after bilateral prophylactic oophorectomy in BRCA1 mutation carriers. *J. Natl Cancer Inst.* **91:** 1475–1479.

Rebbeck, T.R. (2002) Prophylactic oophorectomy in BRCA1 and BRCA2 mutation carriers. *Eur. J. Cancer* **38** (Suppl. 6): S15–S17.

Sgambati, M.T., Stolle, C., Choyke, P.L., Walther, M.M., Zbar, B., Linehan, W.M. and Glenn, G.M. (2000) Mosaicism in von Hippel–Lindau disease: lessons from kindreds with germline mutations identified in offspring with mosaic parents. *Am. J. Hum. Genet.* **66:** 84–91.

Skirton, H. and Eiser, C. (2003) Discovering and addressing the client's lay knowledge – an integral aspect of genetic health care. *Research and Theory for Nursing Practice: an International Journal* **17:** 339–352.

Skirton, H. (2001) The client's perspective of genetic counseling – a grounded theory study. *J. Genet. Counsel.* **10:** 311–329.

Szudek, J., Evans, D.G. and Friedman, J.M. (2003) Patterns of associations of clinical features in neurofibromatosis 1 (NF1). *Hum. Genet.* **112:** 289–297.

Thompson, D., Szabo, C.I., Mangion, J. *et al.* (2002) Evaluation of linkage of breast cancer to the putative BRCA3 locus on chromosome 13q21 in 128 multiple case families from the Breast Cancer Linkage Consortium. *Proc. Natl Acad. Sci. USA* **99:** 827–831.

Winawer, S.J., Fletcher, R, Rex, D. *et al.* (2003) Colorectal cancer screening and surveillance: clinical guidelines and rationale – update based on new evidence. *Gastroenterology* **124:** 544–560.

Wooster, R. and Weber, B.L. (2003) Breast and ovarian cancer. In: *Genomic Medicine* (eds A.E. Guttmacher, F.S. Collins, J.M. Drazen). Johns Hopkins University Press, Baltimore, OH, pp. 118–130.

Worden, J.W. (1991) *Grief Counselling and Grief Therapy.* 2nd Edn. Routledge, London.

Further resources

Alexander, M.F., Fawcett, J.N. and Runciman, P.J. (2000) The patient with cancer. Chapter 31. In: *Nursing Practice Hospital and Home.* 2nd Edn. Churchill Livingstone, London.

Cancer Research UK . [Accessed March 4, 2005] http://www.cancerresearchuk.org/

Cancer Research Institute. [Accessed March 4, 2005] http://www.cancerresearch.org/

Cole, T.R.P. and Sleightholme, H.V. (2000) ABC of colorectal cancer. *BMJ* **321:** 943–946.

National Centre for Biotechnology Information website. [Accessed, March 4, 2005] http://www.ncbi.nlm.nih.gov/books/bv.fcgi

12 Chromosomal and non-traditional patterns of inheritance

12.1 Introduction

It is apparent from looking at family trees that some disorders seem to be inherited although they do not follow a clear Mendelian pattern. There are a variety of mechanisms that could lead to this observation and some of these will be discussed in this chapter.

12.2 Micro-deletion syndromes

> **CASE STUDY – KIM**
>
> A referral is received in the genetics center that concerns a baby (Kim) who has been born with a cleft palate and a heart abnormality. A karyotype has been requested and preliminary cultures have produced a normal result. The referring pediatrician is concerned that further tests should be done. The geneticist sees baby Kim and also talks with the parents. On taking the family tree, she notes that the baby's father, Joseph, has a slight nasal quality to his speech. He reports that he had problems with feeding as a baby and, sometimes, milk used to come out of his nose. The geneticist requests that a specific test is done on a blood sample from the baby, looking for a deletion of chromosome 22 at q11. This test confirms the presence of the deletion and the family is informed and a follow-up appointment made.

22q11 deletion syndrome

This deletion is the most frequent known interstitial deletion in humans and has an incidence of approximately 1 in 4000 live births. It is associated with a large variety of birth defects and the deletion has been linked to a number of diagnoses, including Di George syndrome, and velocardiofacial or Shprintzen syndrome (Scambler, 2000). The clinical features are variable but the main causes of morbidity and mortality are congenital heart defects. In a population study in Atlanta, Georgia, USA, that used multiple sources of case ascertainment, 81% had a major heart defect (predominantly conotruncal abnormalities, such as tetralogy of Fallot, interrupted aortic arch and truncus arteriosus) (Botto *et al.*, 2003). The other features of the syndrome include palatal abnormalities, particularly velopharyngeal incompetence, submucosal cleft palate and cleft palate, characteristic facial features, learning difficulties, immune deficiency, hypocalcalcemia, feeding problems, hearing loss, and renal anomalies.

As follow-up studies of children and adults with the deletion have been conducted, it is becoming apparent that the deletion may also be associated with a particular behavioral and cognitive phenotype, which includes schizophrenia (Murphy, 2002). Although further studies are needed to establish the lifetime risk of developing a serious psychiatric disorder, it is suggested that up to 25% of adults with this deletion may be affected. The deletion is normally only detectable using fluorescent *in situ* hybridization (FISH).

Fluorescent in situ *hybridization*

The method of detecting specific microdeletions or rearrangements using FISH has been described in Chapter 2. This technique is suitable for use in the case described because the health professional has phenotypic evidence that Kim may have a microdeletion of chromosome 22q, and the laboratory are therefore aware of the probes that should be used to examine Kim's chromosome structure at that locus.

CASE STUDY – KIM

At the follow-up appointment, Joseph and his wife are seen by the geneticist with their new baby. The chromosome findings are explained to them and their first questions are about what care baby Kim should receive. He has had surgery on his heart and palate and, apart from some problems post-operatively with his calcium levels, appears to be doing fine. He will require careful medical follow-up and will need his development and growth to be monitored. At some stage, a hearing and speech assessment will be organized and further potential complications will be monitored and evaluated as necessary. At this appointment, the geneticist discusses taking a sample of blood from Joseph and his wife and explains that it is possible that one of them may have the deletion. They are surprisingly reassured by this because, as Joseph says, it has not caused many problems in them. The results come back showing that Joseph also has the deletion and further family studies suggest it arose *de novo* in him.

This may be an explanation for his history of feeding difficulties and palatal problems. Joseph is referred for an evaluation of these, together with a cardiac evaluation. The genetic health professional spends some time discussing with her colleagues how much to tell the family about the risk of learning problems. The family's main concern at the moment is their son's immediate health problems and to what degree he will have learning difficulties. She does raise it as a potential problem with the family but in the context of the wide range of possible problems with this syndrome, emphasizing the variability of the phenotype and the fact that early diagnosis may mean that some of the potential complications can be ameliorated by appropriate diagnosis and treatment.

Recurrence risks

The majority of 22q11 deletions arise *de novo*. In these cases, the recurrence risk is low, although there have been cases of gonadal mosaicism. In such cases, prenatal diagnosis by amniocentesis or chorionic villus biopsy can be offered. In about 7% of cases, the deletion is inherited as in the example above. For Joseph and his wife, the risk of transmitting the deletion would be 50%, but the exact phenotype is difficult to predict. For some families, one option, in addition to karyotyping, is to have detailed high-resolution scanning to detect significant congenital abnormalities to be able to make an informed decision about what action to take during a pregnancy.

The 22q11 deletion may occur as a result of a translocation. This would also give a recurrence risk for a future pregnancy. Studies indicate that many of the *de novo* deletions arise as a result of recombination with unequal crossing-over within the critical region. At the site of the common breakpoints within the region are segments of DNA that consist of low numbers of repeated sections. These variable low-copy-number repeats may predispose to unequal crossing-over during recombination, leading to a deletion and corresponding duplication. This is illustrated in Figure 12.1. The breakpoint of the common 11:22 translocation is also within one of these regions.

Figure 12.1 **Unequal crossing-over**

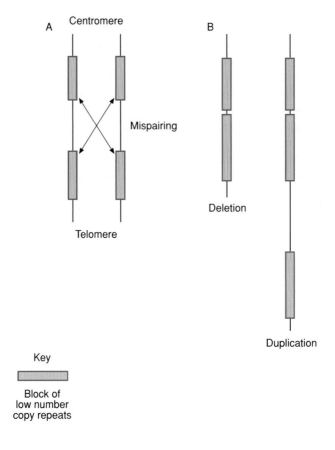

Key

Block of
low number
copy repeats

Other microdeletion syndromes

A number of syndromes are known to be caused by chromosomal
microdeletions, some of which can be seen cytogenetically, but most of which
require FISH analysis. Examples are given in Table 12.1.

As with the 22q deletion, most of these deletions arise *de novo*; however,
they may also be the result of a balanced chromosome rearrangement, a
translocation or a pericentric inversion (see Chapter 2). Parental chromosomes should always be requested if a diagnosis of a microdeletion syndrome is made in a child. If the deletion is not visible cytogenetically, then
the probes that are used are specific to these particular regions; outside of
research settings, it is not possible to do a chromosome-wide scan. The test
is normally requested if the specific syndrome is suspected on clinical
grounds.

Table 12.1 **Examples of microdeletion syndromes**

Syndrome	Chromosome region	Probe target
Cri du Chat syndrome	5p15.2 to the whole of the short arm	Cytogenetically visible
Wolf–Hirschhorn syndrome	4p16.3	Wolf–Hirschhorn critical region
Williams syndrome	7q11.23	Elastin gene
Prader–Willi/Angelman syndrome	15q11.2–q13	PW/AS critical region
Smith–Magenis syndrome	17p11.2	Smith–Magenis critical region
Miller–Dieker syndrome	17p13.3	Lissencephaly gene
Kallman syndrome	Xp22.3	KAL gene
Steroid sulfatase deficiency (X-linked icthyosis)	Xp22.3	Steroid sulfatase gene

12.3 Chromosome abnormalities

In addition to abnormalities of structure, an increase or a decrease in the number of chromosomes leads to well-recognized syndromes, as described in Chapter 2 (Table 2.2).

Down syndrome

Down syndrome is the most common autosomal trisomy with a liveborn incidence of 1.5 per 1000 live births (Jacobs *et al.*, 1992). The features of

Table 12.2 **Risk of a liveborn baby with Down syndrome**

Age of mother at time of baby's birth	Risk of baby affected with Down syndrome
15 years	1 in 1580
20 years	1 in 1530
25 years	1 in 1350
30 years	1 in 901
35 years	1 in 385
37 years	1 in 240
39 years	1 in 145
41 years	1 in 85
43 years	1 in 49
45 years	1 in 28
47 years	1 in 15
49 years	1 in 8

Down syndrome are well known and resources are given at the end of the chapter for further reading. Most cases are as a result of trisomy 21 and this normally occurs because of non disjunction during meiosis, as described in Chapter 2.

Maternal age is a well-known risk factor for Down syndrome, with the risk of having a baby with Down syndrome increasing as maternal age increases. There seems to be no link to paternal age. Table 12.2 shows the risk at birth of a liveborn baby with Down syndrome (Harper, 2004).

Most women are now offered some form of screening test for Down syndrome. As discussed in previous chapters, a screening test is a test that is offered to a whole population or a subset of the population to identify those individuals at high risk and offer them a diagnostic test.

In screening programs for Down syndrome, the diagnostic test will be a chorionic villus sample or amniocentesis. Some centers then proceed to a full karyotype, whereas others might offer a molecular test for the major trisomies (quantitative PCR or interphase FISH).

The choice of tests to use as the screening test is complex. Originally, amniocentesis was offered, using maternal age as the initial screen. Then it was noted that pregnancies affected by Down syndrome showed specific patterns of serum markers. Algorithms were then devised using the markers in combination with maternal age to give an individual risk. More recently, variants in ultrasound findings – such as thickness of the nuchal fold at 10 weeks of pregnancy – have been used. A recent project in the UK investigating the efficacy of a variety of screening strategies illustrates the complexity of the situation (Wald *et al.*, 2003). Table 12.3 shows the variety of screening strategies investigated.

The conclusion of this particular study was that, on the basis of detection rates and safety, the integrated test performed best overall. However, if a nuchal translucency (NT) measurement was not available, the serum integrated test performed best. If a woman did not attend for antenatal care until the second trimester of pregnancy, the quadruple test performed best. For women who chose to have a screening test in the first trimester, the combined test performed best.

This study illustrates the complexity of decision-making in this area, both for providers and purchasers of healthcare and for individual women and their partners. For the practitioner in this situation, counseling regarding risk requires knowledge relating to what the current provision is in his or her area of practice and what the performance of the screening test is.

Table 12.3 Possible screening strategies for Down syndrome

Screening test	Nuchal translucency scan at 10–12 completed weeks of gestation	Serum markers	Gestation (completed weeks)
Integrated test	Yes	PAPPA AFP, uE3, free β-hCG and inhibin-A	10 14–20
Serum integrated test	No	PAPPA AFP, uE3, free β-hCG and inhibin-A	10 14–20
Combined test	Yes	free β-hCG and PAPP-A	10
Quadruple test	No	AFP, uE3, free β-hCG and inhibin-A	14–20
Triple test	No	AFP, uE3 and free β-hCG	14–20
Double test	No	AFP and free β-hCG	14–20
NT measurement	NT	None	12–13

AFP, alphafetoprotein; hCG, total human chorionic gonadotrophin; NT, nuchal translucency; PAPP-A, pregnancy associated plasma protein A; uE3, unconjugated oestriol.

All screening programs have a false-positive rate and a fasle-negative rate. For example, in this context, a false-positive result would involve detecting a woman as high risk who after diagnostic testing proves not to have a Down syndrome pregnancy. A false-negative result would involve reassuring a woman that she is at low risk but missing the fact that she has a Down syndrome pregnancy. Both of these outcomes could potentially cause harm and women who are participating in these programs should be given enough information about these potential outcomes to be able to make an informed decision. Although the consequences of false-positive results in screening programs have been recognized for some time, the potential harm of false-negative results have not been researched so well (Marteau *et al.*, 1992; Petticrew *et al.*, 2000).

If a couple have had a previous child with Down syndrome, before advising parents it is important to check the report of the chromosomal structure of the affected child. If the result shows that the child inherited a full additional copy of chromosome 21 (i.e. 47, XX, +21 or 47, XY, +21), then it is highly likely to have occurred sporadically. In this case, the risk for future pregnancies for this couple will be 1% or twice the mother's age-related risk, whichever is higher (Harper, 2004). This is based on empirical studies and may be an overestimate. If, however, the chromosome result shows that the baby with Down syndrome had a translocation, the parents' chromosomes should be studied to ascertain whether one of them carries a Robertsonian translocation, as this would increase their chances of having a second child with a chromosomal abnormality.

Genomic imprinting

With chromosome disorders such as Down syndrome and 22q11 deletion syndrome, the parental origin of the extra or deleted chromosome does not have an effect on the gross phenotype. Imprinting is a relatively recently described mechanism by which portions of the genome are only activated if they are inherited from a parent of one particular sex (Driscoll *et al.*, 1992). For example, there are families with autosomal dominant glomus tumors (a particular type of non-malignant growth of vascular origin) in which the gene is only expressed if it is inherited from the father. If it is inherited by a male from his mother, he will not express the condition, but his children will have a 50% chance of inheriting the gene and expressing the condition. If his daughters inherit the gene, they will express the condition but their children will not.

Prader–Willi syndrome

As well as single genes being subject to imprinting, whole regions of chromosomes may be subject to the same effect. Prader–Willi syndrome (PWS) – a syndrome that causes excessive appetite, learning difficulties and short stature – is caused by lack of a specific region of the paternally inherited copy of chromosome 15. One of the mechanisms that causes PWS is an imprinting mutation that 'switches off' the region, resulting in failure of expression of critical genes.

Children who have PWS commonly present in the neonatal period with poor sucking reflex, hypotonia and feeding problems. However, in early childhood they start overeating and develop morbid obesity, unless the hyperphagia is externally controlled. They show delayed milestones and have global developmental delay, as well as moderate mental retardation. They also have delayed puberty and hypogonadism.

The condition is caused by loss of the paternal contribution at 15q11.2- 11.3. Some 99% of patients show a parental-specific methylation abnormality and this is usually the first line of testing as it detects abnormal methylation that is caused by deletions, uniparental disomy (UPD) or imprinting mutations. The majority of patients have a deletion that is detectable by FISH, and 10–25% have UPD for the maternal chromosome 15 (Fridman *et al.*, 2000).

Uniparental disomy

Uniparental disomy occurs when an individual inherits both copies of a particular chromosome or chromosome region from one parent rather than one of the pair from the mother and one of the pair from the father.

UPD was first recognized in humans when a child was found to have cystic fibrosis due to the inheritance of two identical copies of chromosome 7 from the mother and none from the father (Spence *et al.*, 1988). For the majority of individuals, when UPD results in the unmasking of a recessive gene in this way, there are no other effects apart from the condition itself. However, as discussed previously, genes are differentially expressed for some chromosomal regions, and this is dependent on whether they are inherited from the mother or from the father.

Trisomy rescue   It is thought that UPD will normally arise through the mechanism of trisomy rescue (Figure 12.2). As is the case in Down syndrome, non disjunction during meiosis will lead to gametes being created that either contain two copies of the chromosome or none. This can happen for any of the chromosomes and leads to the creation of a conception with either three copies of the chromosome or a single copy. It is thought that, in the case of the trisomic conception, a second event may occur, which results in the loss of the extra chromosome. (The cells that have the normal chromosome constitution may have a growth advantage and will be preferentially selected.) If the two remaining chromosomes have been inherited from one parent, then the fetus will have UPD for that chromosome. It is also possible to have UPD for regions of chromosome due to structural abnormalities.

For most chromosomes, there seems to be no phenotypic effect of UPD. Parental-specific imprinting effects have been shown in maternally derived chromosomes 7, 14 and 15, and in paternally derived chromosomes 6, 11, 14 and 15. It is possible that there are also effects from maternally derived chromosomes 2, 16 and 20, and from paternally derived chromosome 20 (Shaffer *et al.*, 2001).

Figure 12.2 **Trisomy rescue resulting in UPD**

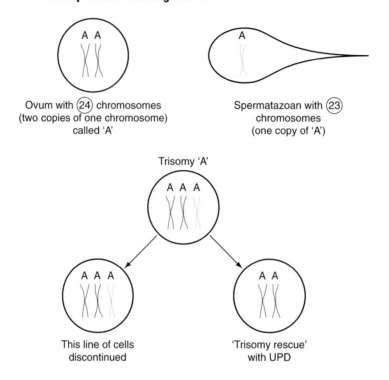

Ovum with (24) chromosomes
(two copies of one chromosome)
called 'A'

Spermatazoan with (23)
chromosomes
(one copy of 'A')

Trisomy 'A'

This line of cells
discontinued

'Trisomy rescue'
with UPD

One situation in which this becomes important is when confined placental mosaicism is detected during chorionic villus sampling (Kalousek and Vekemans, 2000). It is known that the placenta may have mosaic cell lines that are not represented in the constitutional karyotype of the fetus. However, if mosaicism for chromosomes with known imprinted regions is detected, then it is important that further investigations are performed to test for UPD in the fetus.

Angelman syndrome

If there is no maternal contribution of the critical region of chromosome 15, a different syndrome to Prader–Willi results – Angelman syndrome. Angelman syndrome represents one of the best-characterized examples of genomic imprinting in humans and provides an example in which testing strategies and recurrence risks can be discussed in detail.

CASE STUDY – MARCY

Marcy was seen when she was 3 years old. She had been born at term with a normal birth weight. However, her development had been delayed and at the age of 3 years she had no speech, showed a moderate degree of motor delay with ataxia and tremor, and had started to have seizures. Her parents noted that she appeared to be happy and would often laugh for no apparent reason. A diagnosis of Angelman syndrome had been suggested and they wanted to organize testing to confirm the diagnosis.

The features of Angelman syndrome are not present at birth and it can take several years before a diagnosis is made. The symptoms that are usually noticed first are delay of motor milestones and speech. Some 90% of children do walk; however, they typically have a stiff ataxic gait. The behavioral phenotype includes excessive laughter and absent speech. More than 80% of patients develop seizures that have typical EEG changes (Clayton-Smith and Laan, 2003).

Molecular genetics   It is thought that most of the features of Angelman syndrome are a result of the loss of function of the maternally inherited UBE3A allele on chromosome 15. As with Prader–Willi syndrome, there are a variety of mechanisms by which this can occur: deletion, UPD, imprinting defect or mutation, or UBE3A mutations.

The strategy for molecular testing is similar to that of Prader–Willi, however, it will be presented in more detail in this section.

DNA methylation analysis   As with Prader–Willi syndrome, the parental-specific pattern of methylation can be detected in Angelman syndrome. The methylation pattern is determined by using restriction enzymes that differen-tially cut the DNA, depending whether it is methylated or not. By using a probe at a specific locus, a different pattern of bands will be seen, depending on the methylation status of the DNA. At this locus, the DNA is methylated when inherited from the mother and unmethylated when inherited from the father. Normal individuals would have a methylated and unmethylated allele. Individuals with Prader–Will syndrome would only have the maternal allele. Individuals with Angelman syndrome would have the paternal allele.

Approximately 80% of patients with Angelman syndrome have abnormal methylation; this does not distinguish between patients with a deletion, UPD or with an imprinting-center defect.

If the DNA methylation is abnormal, then a deletion is looked for using FISH. If this is normal, then DNA studies will be done looking for UPD (found in 3–7% of patients) or an imprinting center defect. In these situations, if the tests are positive, it is appropriate to perform a high resolution karyotype to check for chromosomal rearrangements that might give a recurrence risk.

If the methylation is normal but Angelman syndrome is strongly suspected, then it is possible to look for mutations in the UBE3A gene. This is done by sequencing the gene, which is labor intensive and expensive but is successful in detecting mutations in 40–50% of patients with normal methylation. In the remaining group of patients with definite Angelman syndrome, the causative mechanism is unknown (Lossie *et al.*, 2001).

The nature of the causative mechanism has implications for the recurrence risk within families, and the information that can be given during genetic counseling will vary.

Table 12.4 indicates the risk of having a future child with Angelman syndrome once an affected child has been diagnosed (adapted from Gene Review on Angelman syndrome, see Gene Review website in 'Web-based resources').

Table 12.4 Recurrence risks in Angelman syndrome

Percentage of families	Genetic mechanism	Risk of another affected child
65–75%	Deletion	<1%
<1%	Inherited structural chromosome abnormality	Could be as high as 50%
3–7%	Paternal UPD	<1%
<1%	Paternal UPD as a result of a Robertsonian translocation	Approaching 100% if father has a balanced 15:15 translocation
0.5%	Deletion of imprinting center	Could be 50% if mother also has deletion
2.5%	Imprinting defect without deletion	<1%
11%	UBE3A mutation	Could be 50% if mother also has mutation
10–15%	No abnormality identified	Most cases are not familial; recurrence risk not known but could be 50%

If a UBE3A mutation, an imprinting-center deletion or a structural chromosomal arrangement is identified, then other family members may be at risk.

The process of setting the imprinting pattern occurs during gametogenesis and the pattern is reset at every generation. For example, a woman has a

---

CASE STUDY – MARCY

Marcy's genetic test indicates that she has a deletion of the imprinting center, which has been inherited from her mother. This means that relatives of the mother may be at risk of having a child with Angelman syndrome. The situation is difficult to explain and Marcy's mother is seen on a couple of occasions so that she can absorb the information. The people that are at risk are her sisters and her brother's daughters, if the deletion is passed on by her brother. As the genetics practitioner explains, if the deletion had been inherited from Marcy's grandfather it would not cause Angelman syndrome in Marcy's mother or her brothers and sisters. This is because the methylation pattern on the chromosome from Marcy's grandmother would be correct and they would have a maternal contribution. However, if any female with the mutation passes the deletion on, the methylation pattern would be incorrect, there would be no maternal contribution and the child would have Angelman syndrome. So, for Marcy's uncle, if he inherited the deletion from his father he would be fine, and if his daughters inherited it from him they would be fine. However, if their children inherited it they would have Angelman syndrome. Figure 12.3 outlines the details of Marcy's family tree.

Figure 12.3 **Marcy's family tree**

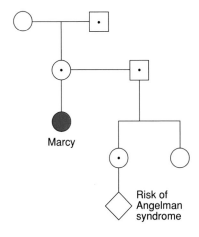

211

paternal and maternal copy of every chromosome with the appropriate imprinting pattern. However, when gametes are produced the imprinting pattern of all the chromosomes will be reset to a maternal pattern. The phenomenon of imprinting may be an explanation for the high embryo and fetal loss rates in studies in which mammalian embryos have been cloned. In cloned embryos, the chromosomes will not be correctly imprinted as they have not been through the process of gametogenesis. As well as the ethical objections, this is further argument against human cloning (Human Genetics Advisory Commission, 1998).

12.4 Mitochondrial inheritance

Mendelian patterns of inheritance relate to alterations in the DNA that form the chromosomes in the nucleus of the cell. An additional important cause of human disease is mitochondrial mutations.

The mitochondria are organelles within cells that have their own genome. A single cell will contain many mitochondria. The mitochondrial genome in humans is double-stranded DNA and is self-replicating. It encodes 13 proteins – all of which are involved in respiratory-chain complexes, therefore having a role in the energy production of the cell – and a number of transfer and ribosomal RNAs (Thorburn and Dahl, 2001).The two unusual features regarding the pattern of inheritance (see Figure 12.4) that are associated with mitochondrial mutations are:

(1) the disease is inherited through the maternal line only (matrilineal);
(2) there is considerable variability in how the disease is manifested.

Inheritance is matrilineal because only the ovum contributes mitochondria to the developing embryo (the zygote). It is assumed that, apart from in very rare situations, sperm do not contribute mitochondria as the mitochondria in the sperm are carried in the tail, which does not penetrate the ovum at fertilization. Therefore, a mitochondrial condition can affect both sexes but will only be inherited from an affected mother.

Some nuclear genes also have a role in mitochondrial function but they would follow normal Mendelian patterns of inheritance and are not discussed here.

The variability of the condition is partly because the DNA sequence in each individual mitochondrium within a cell may not be identical. In some mitochondrial diseases, all the mitochondrial genomes may be identical and

Figure 12.4 Family tree showing maternal inheritance and variable phenotype

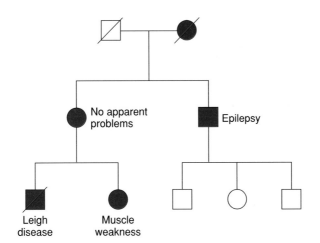

may all carry the causative mutation (i.e. homoplasmy). In other cases, there may be a mixed population of normal and mutant genomes (i.e. heteroplasmy).

Mitochondrial heteroplasmy can be transmitted from mother to child but there can be considerable variation in the proportion of mutant and normal mitochondrial genomes. For example, a woman may have a child who presents with a severe encephalopathy and then dies. The child is subsequently found, in a muscle biopsy and in blood tests, to have a mitochondrial mutation. When the mother is tested, she is shown to have no mutation in her blood but a muscle biopsy does demonstrate that the mutation exists at a low level. However, for future pregnancies, all that can be said is that the child will inherit the mutation, but the level of mutation cannot be determined.

Mitochondria are within every cell in the body; every tissue in the body could be affected. However, certain constellations of symptoms and signs do occur together and allow for specific diagnoses to be made. An additional complication is that particular mitochondrial mutations may be associated with a number of these diagnoses and are not specific.

Some disorders predominantly affect one system, such as Leber hereditary optic neuropathy (LHON), which affects the eye. Many involve multiple body systems and often present with predominantly neurological or neuromuscular symptoms. Examples of these are chronic progressive external ophthalmoplegia (CPEO), mitochondrial encephalomyopathy with lactic acidosis and stroke-like episodes (MELAS), myoclonic epilepsy with ragged-red fibers (MERRF), neurogenic weakness with ataxia and retinitis pigmentosa (NARP), or Leigh syndrome (LS). However, there is considerable overlap and many patients do not fit into one particular category (DiMauro and Schon, 2001).

Genetic counseling in mitochondrial disorders

For families with a mitochondrial disorder, an ideal situation would involve being able to give precise recurrence risks and accurate information about the potential phenotype, together with reproductive options. This is not currently possible for the majority of cases.

Males will not pass on mitochondrial disorders to their children (unless the disorder is caused by a nuclear-encoded gene); therefore, the risk of recurrence is for women. It is possible to detect mitochondrial mutations in prenatal samples. However, the percentage of mutated mitochondria in the fetal sample may not reflect the percentage in tissues at birth. For certain specific mutations that cause NARP or Leigh disease, studies have been conducted that have attempted to calculate the probability of having a severely affected child based on the percentage of mutated mitochondria in the mother. However, the findings give very broad ranges for their estimates and need to be used with caution (Thorburn and Dahl, 2001).

Mitochondrial disorders are an example of conditions in which genetics counseling is really focused on helping people reach decisions that are congruent with their values and beliefs under conditions of considerable uncertainty. As such, they provide a challenge for genetics clinicians, but also an opportunity through counseling to assist people in coming to decisions that are right for them.

12.5 Conclusion

In this chapter, a variety of mechanisms of inheritance have been presented. The complexity of genetic information is likely to increase and will require the professional working in specialist genetic healthcare to continuously update their knowledge to practice competently.

Study questions

1. Read Marcy's case study again.

Task A

Write down a testing strategy for Marcy. Find out what tests are available locally to you, where you would send samples and how long it would take to get results.

Task B

Angelman syndrome and Prader–Willi syndrome are caused by incorrect imprinting of the same region of chromosome 15. Think of a mechanism whereby you could get both syndromes occurring in the same family and write down a hypothetical family tree.

2. A woman is referred to you because she has a brother with Down syndrome.

(a) Note down what issues you may want to discuss with her during counseling.
(b) What (if any) investigations would you want to know the results of?
(c) What are her reproductive options?

References

Botto, L.D., May, K., Fernhoff, P.M. *et al.* (2003) A population-based study of the 22q11.2 deletion: phenotype, incidence, and contribution to major birth defects in the population. *Pediatrics* **112**: 101–107.

Clayton-Smith, J. and Laan, L. (2003) Angelman syndrome: a review of the clinical and genetic aspects. *J. Med. Genet.* **40**: 87–95.

DiMauro, S. and Schon, E.A. (2001) Mitochondrial DNA mutations in human disease. *Am. J. Med. Genet.* **106**: 18–26.

Driscoll, D., Waters, M., Williams, C.A., Zori, R., Glenn, C.C., Avidano, K.M. and Nicholls, R. (1992) A DNA methylation imprint, determined by the sex of the parent, distinguishes the Angelman and Prader–Willi syndromes. *Genomics* **13**: 917–924.

Fridman, C., Varela, M.C., Kok, F., Setian, N. and Koiffmann, C.P. (2000) Prader–Willi syndrome: genetic tests and clinical findings. *Genet. Test.* **4**: 387–392.

Harper, P.S. (2004) *Practical Genetic Counselling.* 6th Edn. Oxford University Press, Oxford.

Human Genetics Advisory Commission. Cloning issues in reproductive science and medicine (1998) [Accessed August 4, 2004] http://www.advisorybodies.doh.gov.uk/hgac/papers/papers_d.htm

Jacobs, P., Browne, C., Gregson, N., Joyce, C. and White, H. (1992) Estimates of the frequency of chromosome abnormalities detectable in unselected newborns using moderate levels of banding. *J. Med. Genet.* **29**: 103–108.

Kalousek, D.K. and Vekemans, M. (2000) Confined placental mosaicism and genomic imprinting. *Best Pract. Res. Clin. Obstet. Gynaecol.* **14**: 723–730.

Lossie, A.C., Whitney, M.M., Amidon, D. *et al.* (2001) Distinct phenotypes distinguish the molecular classes of Angelman syndrome. *J. Med. Genet.* **38**: 834–845.

Marteau, T.M., Cook, R., Kidd, J., Michie, S., Johnston, M., Slack, J. and Shaw, R.W. (1992) The psychological effects of false-positive results in prenatal screening for fetal abnormality: a prospective study. *Prenat. Diagn.* **12**: 205–214.

Murphy, K.C. (2002) Schizophrenia and velo-cardio-facial syndrome. *Lancet* **359**: 426–430.

Petticrew, M.P., Sowden, A.J., Lister-Sharp, D. and Wright, K. (2000) False-negative results in screening prgrammes:systematic review of impact and implications. *Health Technol. Assess.* **4**: no 5.

Scambler, P.J. (2000) The 22q11 deletion syndromes. *Hum. Mol. Genet.* **9**: 2421–2426.

Shaffer, L.G., Agan, N., Goldberg, J.D., Ledbetter, D.H., Longshore, J.W. and Cassidy, S.B. (2001) American College of Medical Genetics statement of diagnostic testing for uniparental disomy. *Genet. Med.* **3**: 206–211.

Spence, J.E., Perciaccante, R.G., Greig, G.M., Willard, H.F., Ledbetter, D.H., Hejtmancik, J.F., Pollack, M.S., O'Brien, W.E. and Beaudet, A.L. (1988) Uniparental disomy as a mechanism for human genetic disease. *Am. J. Hum. Genet.* **42**: 217–226.

Thorburn, D.R. and Dahl, H.H. (2001) Mitochondrial disorders: genetics, counseling, prenatal diagnosis and reproductive options. *Am. J Med. Genet.* **106**: 102–114.

Wald, N.J., Rodeck, C.H., Hackshaw, A.K., Walters, J., Chitty, L. and Mackinson, A.M. (2003) First and second trimester antenatal screening for Down's syndrome: the results of the serum, urine and ultrasound screening study (SURUSS). *Health Technol. Assess.* **7**: no 11.

Further resources

Assert – Angelman support education and research trust. [Accessed March 3, 2005] http://www.angelmanuk.org/

Gene Reviews. [Accessed March 3, 2005] http://www.geneclinics.org

United Mitochondrial Disease Foundation. [Accessed March 3, 2005] http://www.umdf.org/

13 Multifactorial inheritance and common diseases

13.1 Introduction

Mendelian inheritance patterns observed in families are due to the action of single highly penetrant genes. Before the identification of genes, the observation of patterns in families was based on the fact that the phenotype was assumed to be either absent or present (e.g. an individual has an extra finger or a normal number of fingers, has phenylketonuria or does not have phenylketonuria).

This is an oversimplification even for Mendelian disorders, as illustrated by the case of thalassemia in which the phenotype shows considerable variability based on interactions between genes and environmental factors (Weatherall, 2000). An observable phenotype is rarely the consequence of the action of one DNA variant, but is the consequence of a pathway that may include many genes, proteins and other environmental mediators. For example, a person's final height will be determined by the combination of genes that they inherit from their parents; their growth and development in fetal life; and their nutrition, health, and growth and development during childhood.

Common birth defects, such as neural-tube defects, cleft lip and palate, and congenital heart disease, rarely follow Mendelian patterns of inheritance, but clearly genetic factors are involved as there are more frequent recurrences within families than would be expected due to chance. The formation of a normal lip and palate is the end result of a complex developmental pathway. If enough genetic and environmental factors are present, then the pathway can be disrupted and clefting occurs.

In one case, the factors may be more genetic than environmental; in another, environmental influences may predominate. However, if enough

Figure 13.1 Threshold effect

of these small individual effects accumulate, a threshold is reached at which the developmental pathway is disrupted. The concept of a threshold effect is used to explain the presence or absence of a characteristic such as cleft lip and palate (Figure 13.1).

In genetics counseling for multifactorial conditions, the risk that is given is not based on the theoretical risk derived from Mendelian theory, but on empirical risks, which are derived from large observational surveys.

13.2 Neural tube defects

Neural tube defects and counseling

The neural tube includes the brain and the spinal cord. This develops early in fetal life, and interruption to closure of the tube causes both spina bifida and anencephaly. Closure of the neural tube begins in the area of the cervical spine, with the spinal column forming distally in the direction to the coccyx, while the cranium closes over the brain. Failure of this process results in a neural-tube defect. This term includes spina bifida, anencephaly and any other defect that is a failure of closure of the neural tube.

> **CASE STUDY – JODY**
>
> Jody has spina bifida and is seeing a healthcare provider for advice before starting a family. She has a number of concerns, some of which relate to her physical ability to get pregnant, carry a baby to full term and deliver the baby safely. Her other concerns relate to the risk of any baby also having spina bifida. The health worker organizes referral to an obstetrician so that Jody can be assessed and discusses with her the risks to this pregnancy.
>
> She had been seen by a geneticist when she was young and at that stage there was no other family history of neural-tube defects and Jody was carefully examined to see if she had any other congenital abnormalities. There was no relevant history and it appeared that, for Jody, the spina bifida was an isolated event.
>
> The health worker spent some time exploring with Jody what she felt about having a baby like herself, would it be something she would want tested for in a pregnancy, and if she had a test what would she do if the result showed the baby had a similar degree of disability as Jody or perhaps would be more severely disabled. They also talked about preventative things that Jody could do, including taking high-dose folic acid.

Hydrocephalus is often associated with neural tube defects as a secondary consequence of disturbance of the circulation of the cerebrospinal fluid. There is a rare form of X-linked hydrocephalus associated with mutations in the LCAM gene and hydrocephalus may also be part of other rare syndromes. Careful examination is needed to exclude neural tube defects before considering that a baby with hydrocephalus has one of these rare isolated forms.

The incidence of spina bifida varies between countries, regions of countries and also over time. A public-health campaign in Australia demonstrated a reduction in the total number of births with a neural tube defect from around 2 children in every 1000 in 1996 to approximately 1.1 in every 1000 in 1999. As spina bifida accounts for approximately half of all neural tube defects, this means the birth rate for spina bifida since 1996 has been between 0.5 and 0.65 per 1000 births (Chan *et al.*, 2001).

In the absence of other abnormalities in the fetus, neural tube defects are not inherited as a Mendelian disorder – that is, they do not follow one of the known patterns of inheritance. However, empirical data show that couples who have one child with a neural tube defect are at greater risk of having a second child with a neural tube defect than other couples in the general population. Following the diagnosis of one child with a neural tube defect, the risk of recurrence in each subsequent pregnancy for that couple is about 4%. However, the risk can be lowered dramatically (to about 1%) by taking maternal folic acid supplements (Harper, 2004).

In 1991, a randomized controlled trial of the effect of maternal supplements of folic acid on the rate of neural tube defects in their offspring was reported (Wald, 1991). The study was terminated early, as it became clear that folic acid was having an effect in reducing the number of children born with spina bifida. It was therefore unethical to continue to place women in the control group, as this was knowingly exposing the fetus of each of those women to greater risk.

Since that time, it has been recommended that women who may become pregnant take daily folic acid for at least 2 months before conception, and for the first 12 weeks of pregnancy. The dose for women whose risk of having a child with a neural tube defect is considered to be at the population level is about 0.4 mg daily, but women whose children appear to be at higher risk are advised to take 4–5 mg per day.

Folic acid supplements are advised to be taken before conception because it takes some weeks to raise the maternal levels, and because the neural tube has already started to form before the mother is aware of her pregnancy. This means, of course, that some women will be taking folic acid for many months, even years, before conception occurs. However, as a B-group vitamin, excess folic acid is excreted in the urine, and toxic levels do not therefore occur in the mother.

Although dietary advice is important, increasing the levels of folic acid in the maternal diet is not considered sufficient, as the prevalence of neural tube defects does not appear to vary significantly with maternal diet. It may be that some women do not absorb folic acid from the diet as well as others.

Prospective mothers who have had a previous child with a neural tube defect or who are on anti-epileptic medication should be prescribed the higher dose of folic acid because of their higher risk of having a child with a neural tube defect.

When seeing a family for genetics counseling, it is important to distinguish between cases in which the neural tube defect is isolated and those in which it forms part of a syndrome. It should be noted that spina bifida and anencephaly are part of the same spectrum and the recurrence risk includes the whole spectrum. This may make a difference to how a family will perceive the recurrence risk.

MTHFR *gene and neural tube defects*

The finding that folate supplementation led to a reduction in neural tube defects raises interest in the possibility that genes in the folate metabolism pathway might have a role to play in either predisposition to or protection against neural tube defects. One candidate gene was methylenetetra-hydrofolate reductase (MTHFR). Deficiencies in MTHFR with a specific genotype (TT) result in highly elevated plasma homocysteine and reduced folate. Data from association studies gave conflicting results when looking for an association between neural tube defects and the genotype in the affected individual.

These studies have been criticized on methodological grounds (Posey *et al.*, 1996). Many of them did not include appropriate control groups and did not account for potential interactions between intake of dietary folate and maternal genotype, or for an effect of maternal genotype directly. The role of this gene in the genesis of neural tube defects is unclear.

The process of neural tube formation and subsequent differentiation is complex and will be under the control of many different genetic and biochemical pathways. Any individual genetic factor may have a small effect and will therefore be difficult to assess. In public-health terms, the implementation of folic acid supplementation has been a success in reducing the incidence of neural tube defects, regardless of the genetic mechanisms that might be involved.

13.3 Genetics of common diseases

As illustrated in the example of neural tube defects and the MTHFR gene, the pathway from phenotype to genotype is complex and will require investigation of well-designed and methodologically sound studies. Recent developments in genetics, computing, and bioinformatics are making it possible to do large enough studies to elucidate associations between genetic variation and complex human phenotypes. These tools are being applied in large-scale epidemiological studies in order to dissect out the genetic component of complex human diseases, such as cancer, diabetes, and coronary heart disease.

Many studies have been done in high-risk families or selected groups and have identified genetic variants that contribute to these diseases. These studies have had the advantage that they focus on one phenotype – for example, asthma – and look for areas of genetic variation that people with asthma have in common. Once an area is identified, the search for candidate genes can begin. This approach has been successful in identifying the major disease genes that are responsible for classic genetic diseases, such as Huntington disease, cystic fibrosis, and dominantly inherited breast cancer.

As a result of these sorts of techniques, genetic factors have been implicated in many different conditions, such as cancer, diabetes, asthma, and Alzheimer disease. Some examples will be discussed below.

Folate metabolism and colon cancer

In the same way that epidemiological evidence suggested that folate was involved in the genesis of neural tube defects, there is evidence to suggest that folate has a role in the etiology of colorectal carcinomas and adenomas (Little and Sharp, 2002). It is suggested that folate deficiency predisposes to cancer by causing DNA hypomethylation and oncogene activation or by affecting DNA synthesis, which results in failure of DNA repair.

A recent review of five of the genes involved in folate metabolism evaluates the role of polymorphisms of those genes as risk factors for colon cancer (Sharp and Little, 2004). The conclusion of this review is that a specific polymorphism in the MTHFR gene (C677T) is associated with a reduction in colon cancer risk. In interaction studies, it was shown that high folate intake and alcohol both appeared to increase the protective effect. This allele is associated with reduced enzyme activity, which is opposite to what might be expected. As a result of these findings, the authors suggest that investigators now focus on the role of this gene in DNA synthesis. There were no significant findings with any of the other polymorphic genes.

It is too early to promote intervention based on either phenotypic or genotypic testing for genes and metabolites involved in the folate pathway as the roles of folate-metabolizing genes, folate and dietary factors are complex. The additional problem with research of this type is that methodologies are currently not available to clarify the role of the various factors and their multiple interactions.

Alzheimer disease

Alzheimer disease is the most common form of dementia in North America and Europe. The prevalence increases with age, with estimates that approximately 50% of people aged over 85 years will be affected. Families showing autosomal dominant inheritance with early onset have been observed for many years. However, they account for less than 2% of cases. There seems to be no clinical or neuropathological differences between familial and sporadic cases. It has also been recognized that siblings of individuals with Alzheimer disease are at increased risk and that recurrence is more frequent in monozygotic compared with dizygotic twins, suggesting a genetic component.

Although there are no proven environmental factors that are associated with Alzheimer disease, some medical conditions are known to be risk factors. Individuals with Down syndrome are at risk of developing early-onset dementia. This observation led to the identification of the amyloid precursor protein gene on chromosome 21, which is also responsible for some familial cases. Heart disease and its medical risk factors, such as hypertension, arterial disease, and hypercholesterolemia, may also predispose to Alzheimer disease.

Including the gene on chromosome 21, three genes have been identified that cause autosomal dominant Alzheimer disease: presenilin-1 on chromosome 14 and presenilin-2 on chromosome 1. Predictive testing is available for individuals who are at a 50% risk of developing Alzheimer disease because a mutation has been identified in their family. In this situation, the counseling is similar to that for Huntington disease as the test is for an early-onset neurological disorder for which there is no treatment.

Genes that are associated with non-Mendelian disease are shown in Table 13.1 (Bertoli-Avella *et al.*, 2004).

Table 13.1 Genes associated with non-Mendelian Alzheimer disease

Chromosome	Gene
19q13.2	Apolipoprotein (APOE)
12p13	Not known
10p13	Not known
9p	Not known
2q,15q,20p	Reported in single data sets, but not confirmed

Source: Strachan and Read, Human Molecular Genetics, Third Edition. © 2004, Garland Science.

APOE4 and Alzheimer disease   There are three common variants of the Apolipoprotein (APOE) gene: ε2, ε3 and ε4. The gene has been extensively investigated because of its role in lipid metabolism and ischemic heart disease. Mortality from ischemic heart disease is related to the ε4 allele (Eichner *et al.*, 1993).

The most common allele is ε3, which occurs in 60–80% of humans. Studies have shown an association between possession of the ε4 allele and Alzheimer disease and a possible protective effect of the ε2 allele. It is suggested that the effect of the ε4 allele is to lower the age of onset of the disease. It has been estimated that there will be a difference of age of onset of 17 years between a person who is ε4ε4 and a person who is ε2ε3 (Warwick *et al.*, 2000).

This association has been consistently replicated and may provide a basis for research into the natural history of the disease and a target for therapeutic intervention. However, caution has been expressed regarding using this in clinical practice, and genetic testing for APOE alleles is not recommended. It does not provide the accuracy that would be required for a clinical test. Alzheimer disease develops in the absence of ε4 alleles and many individuals with ε4alleles will not develop Alzheimer disease (ACMG ASHG, 1995).

In this case, the variant that was being investigated was originally associated with one phenotype – cardiovascular disease. It was only later recognized that this genotype was also associated with a different phenotype. The simplistic model of one-gene–one-disease that may be appropriate when considering Mendelian disorders is not relevant when considering more common disorders in which the etiology is complex. It may be 'many-genes–one-disease' or 'one-gene–many-diseases'.

This will become more important to consider when multiple gene variant testing is introduced in large epidemiological studies. A variant that is assumed to be associated with one particular phenotype may later be found to be associated with a different phenotype, which may have much more serious consequences.

Cardiovascular disease

Cardiovascular disease is the leading cause of morbidity and mortality in adults in Western societies. Decades of research have provided epidemiological evidence for numerous physiological, environmental and behavioral factors that influence the risk of developing cardiovascular disease (Grundy *et al.*, 1998). These are well known and include smoking, obesity and

Figure 13.2 Model of gene–environment interaction

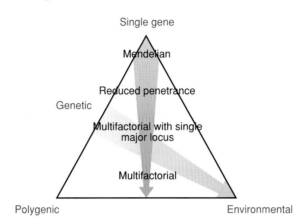

lack of exercise. However, it is clear from personal experience that some people who lead extremely unhealthy lives live to an healthy old age and others who eat healthily, do not smoke, and take regular exercise still die of coronary disease at a young age.

The development of cardiovascular disease is mediated both by direct genetic and environmental effects and also intermediate traits, which are themselves mediated by genetic and environmental effects (Figure 13.2). In addition, behavioral and psychological factors, such as smoking behavior, feed into the interactions.

Familial hypercholesterolemia

As with Alzheimer disease, there is a subset of cardiovascular disease that is caused by single genes. Familial hypercholesterolemia is one of the most studied of these.

Familial hypercholesterolemia results primarily from mutations in the low-density lipoprotein receptor (LDLR) gene, resulting in malfunction and thus leading to very high levels of cholesterol. These high cholesterol levels result in early and severe atherosclerosis and high mortality from coronary heart disease. Early detection is possible and can be beneficial because treatment with statins reduces cholesterol levels and improves clinical outcomes. The disorder is autosomal dominant and, once a proband has been identified, it is important to consider offering testing to at-risk family members (Marks *et al.*, 2000).

Studies of the function of the LDLR gene helped delineate the important role of cholesterol metabolism in cardiovascular disease. However, monogenic disorders only account for approximately 5% of middle-aged patients who have cardiovascular disease.

Numerous studies have been conducted to attempt to delineate other genes that are associated with cardiovascular diseases. As is the case in previous examples, studies produce conflicting results and are limited by the lack of large population samples and appropriate methodologies. Current strategies for preventing morbidity and mortality associated with cardiovascular disease focus on targeting health-promotion advice to all individuals and identifying people with the intermediate factors, such as high blood pressure or high cholesterol, and then treating those.

Asthma

Asthma is an example of an extremely complex disorder in which there are probably multiple genetic and environmental risk factors that have many complex interactions. Unlike cardiovascular disease, there are no subsets of families with single gene disorders and the potential risk factors are not well established. Asthma is a clinical condition that consists of a cluster of related disorders and will have no single etiology.

The prevalence of asthma has increased in the last 20 years and, undoubtedly, environmental risk factors have a central role in any one individual developing this clinical entity. However, twin and family studies have consistently suggested that important genes, in interaction with environmental factors, underlie the predisposition to both allergy and asthma (Patino and Martinez, 2001).

Advances in the understanding of the molecular biology of asthma have identified a large number of candidate genes of interest. These include genes that are involved in inflammation and immune regulation, genes that regulate structural components of the lung, and many others (Anderson and Cookson, 1999).

Asthma susceptibility genes are being identified, but the research is in an early stage and faces many problems, which will be the same for other common clinical conditions. These include the need for standard definitions of asthma phenotypes and characterization of the intermediate phenotypes, well-defined studies in well-characterized populations, and the methodological capability to determine complex interactions.

13.4 Predisposition genes

Using genetic factors as predictors of disease has until now only been applicable to disorders that are caused by single, highly penetrant genes. In many of these disorders, the prediction of disease has not led to prevention or treatment. In the future, this may change.

It is suggested that advances arising from the Human Genome Project will lead to an era of personalized medicine in which individuals are identified who are at risk of diseases, allowing for focused screening, early detection, and individualized treatment (Collins and McKusick, 2001). However, to translate these advances into health benefits requires much work.

The scenario of prediction and prevention is based on a series of assumptions.

- There are genetic risk factors and the risks can be quantified.
- People are willing to be tested.
- There is effective prevention or therapy.
- People will accept the treatment or therapy.

All the above require rigorous evaluation before any technology can be introduced into the healthcare arena. It is recognized that large population-based studies are needed to try to establish answers to the first area – the relationship between genetic factors and disease.

In the UK, the research councils, charities and government propose setting up a large DNA database (UK Biobank) and linking it to medical records in order to facilitate such studies. At the current time, the scientific advances from these types of endeavors have yet to be translated into treatment or prevention strategies that improve health and that may be many years away.

13.5 Pharmacogenetics

One area in which advances may move into clinical practice sooner is the area of pharmacogenetics. The aim of pharmacogenetics is to improve the therapeutic efficiency of drugs and to reduce toxicity.

Traditionally, medicines were developed to treat signs and symptoms of a condition. Where different medications were available, the patient tried these until a satisfactory option was found. This, of course, means that a proportion of patients were being treated with drugs that had no beneficial effect for them, or that produced unwanted side-effects or complications.

Testing a patient for genetic variation before prescribing a drug would enable the right drug to be administered in the right dose. A simple genetic test to look at the polymorphic variation present in the patient would provide the information needed to target the drug (Nicol, 2003). For example, some patients have difficulty metabolizing beta receptor antagonists, because of lack of function of the cytochrome P450 enzymes. Using pharmacogenetics, these patients could be identified and toxicity from the drug avoided.

However, pharmacogenetic research faces the same difficulties and problems as research into genetic risk factors, and using individual genetic variation to determine therapy is not yet in clinical practice. Genetic markers have been used to define some genetic subtypes of disease to target therapy. For example, in hepatitis C, in which the viral genotype is used to tailor treatment, and in breast cancer in which tumors that are positive for Her2 respond to a specific drug, Herceptin (Melzer et al., 2003).

It is possible that pharmacogenetics is one area in which genetic science will contribute to treatment in the near future.

13.6 Conclusion

The genetic contribution to common diseases is complex and the phenotype will be the result of a chain of environmental and genetic effects and interactions. The genetic professional needs be able to evaluate the utility of research findings and understand the complexity in order to translate advances in knowledge into patient benefit.

Study questions

You receive the following referral letter:

Dear Practitioner

Re: Martin Doe

I have been Martin's family doctor for many years and he is becoming increasingly anxious about his family history of dementia. His mother died after many years in a nursing home and her brother has recently been admitted for long-term care. Martin's grandfather also has dementia but is still at home being cared for by his family.

Task A

What information do you need from Martin?

Task B

Draw two hypothetical family trees; one giving Martin a high risk, one a lower risk.

Task C

What counseling issues might be important in each situation?

References

ACMG ASHG (1995) Statement on use of apolipoprotein E testing for Alzheimer disease. American College of Medical Genetics/American Society of Human Genetics Working Group on ApoE and Alzheimer disease. *J. Am. Med. Assoc.* **274**: 1627–1629.

Anderson, G.G. and Cookson, W.O. (1999) Recent advances in the genetics of allergy and asthma. *Mol. Med. Today* **5**: 264–273.

Bertoli-Avella, A.M., Oostra, B.A. and Heutink, P. (2004) Chasing genes in Alzheimer's and Parkinson's disease. *Hum. Genet.* **114**: 413–438.

Chan, A., Pickering, J., Haan, E., Netting, M., Burford, A., Johnson, A. and Keane, R.J. (2001) Folate before pregnancy: the impact on women and health professionals of a population-based health promotion campaign in South Australia. *Med. J. Aust.* **174**: 631–636.

Collins, F.S. and McKusick, V.A. (2001) Implications of the Human Genome Project for medical science. *J. Am. Med. Assoc.* **285**: 540–544.

Eichner, J.E., Kuller, L.H., Orchard, T.J., Grandits, G.A., McCallum, L.M., Ferrell, R.E. and Neaton, J.D. (1993) Relation of apolipoprotein E phenotype to myocardial infarction and mortality from coronary artery disease. *Am. J. Cardiol.* **71**: 160–165.

Grundy, S.M., Balady, G.J., Criqui, M.H. *et al.* (1998) Primary prevention of coronary heart disease: guidance from Framingham: a statement for healthcare professionals from the AHA Task Force on Risk Reduction. American Heart Association. *Circulation* **97**: 1876–1887.

Harper, P.S. (2004) *Practical Genetic Counselling.* 6th Edn. Oxford University Press, Oxford.

Little, J. and Sharp, L. (2002) Colorectal neoplasia and genetic polymorphisms associated with folate metabolism. *Europ. J. Cancer Prev.* **11**: 105–110.

Marks, D., Wonderling, D., Thorogood, M., Lambert, H., Humphries, S.E. and Neil, H.A.W. (2000) Screening for hypercholesterolaemia versus case finding for familial hypercholesterolaemia: a systematic review and cost-effectiveness analysis. *Health Technol. Assess.* **4**: no 29.

Melzer, D., Raven, A., Detmer, J., Ling, T. and Zimmern, R. *My Very Own Medicine. What Must I Know?* (2003) Department of Public Health and Primary Care, University of Cambridge, Cambridge.

Nicol, M.J. (2003) The variation of response to pharmacotherapy: pharmacogenetics – a new perspective to 'the right drug for the right person'. *Medsurg. Nurs. J. Adult Health* **12**: 242–249.

Patino, C.M. and Martinez, F.D. (2001) Interactions between genes and environment in the development of asthma. *Allergy* **56**: 279–286.

Posey, D., Khoury, M., Mulinare, J., Adams, J. and Ou, C.Y. (1996) Is mutated MTHFR a risk factor for neural tube defects? *Lancet* **347**: 686–687.

Sharp, L. and Little, J. (2004) Polymorphisms in genes involved in folate metabolism and colorectal neoplasia: a HuGE review. *Am. J. Epidemiol.* **159**: 423–443.

Wald, N. (1991) Prevention of neural-tube defects – results of the Medical Research Council Vitamin Study. *Lancet* **338**: 131–137.

Warwick, D.E., Payami, H., Nemens, E.J., Nochlin, D., Bird, T.D., Schellenberg, G.D. and Wijsman, E.M. (2000) The number of trait loci in late-onset Alzheimer disease. *Am. J. Hum. Genet.* **66**: 196–204.

Weatherall, D.J. (2000) Science, medicine, and the future: single gene disorders or complex traits: lessons from the thalassemias and other monogenic diseases. *BMJ* **321**: 1117–1120.

Further resources

Human Genome Project Information [Accessed March 3, 2005] http://www.ornl.gov/sci/techresources/Human_Genome/publicat/hgn/hgn.shtml

Khoury, M., Little, J. and Burke, W. (2004) *Human Genome Epidemiology.* Oxford University Press, New York.

McLeod, H.L. and Evans, W.E. (2001) Pharmacogenomics: unlocking the human genome for better drug therapy. *Ann. Rev. Pharmacol. Toxicol.* **41**: 101–121.

My very own medicine: a report of a project on pharmacogenetics funded by the Wellcome Trust. [Accessed March 3, 2005] http://www.phgu.org.uk/about-phgu/pharmacogenetics.html

http://www.phgu.org.uk/about_phgu/HD1032%20CHR_MySery%20own%20Medic.pdf

National Birth Defects Prevention Network. [Accessed March 3, 2005] www.nbdpn.org/NBDPN

Roses, A.D. (2000) Pharmacogenetics and the practice of medicine. *Nature* **405**: 857–865.

UK Biobank. [Accessed March 3, 2005] http://www.ukbiobank.ac.uk/

Appendix I

Use of genetics knowledge by advanced-level practitioners

Oncology and cancer services

Role of the practitioner

Nurses and counselors working with clients who have suspected or confirmed cancer provide care for clients through their cancer journey, supporting them through their diagnosis, treatment and follow-up.

Case examples

Reassuring those at low genetic risk   A 51-year-old man with colon cancer is very concerned about his children because his maternal uncle also developed the disease at the age of 74 years. There is no other family history of cancer.

Recognizing when family history is significant   A 45-year-old woman who has breast cancer has just discovered that her paternal aunt died from ovarian cancer at the age of 50 years, not cervical cancer as was previously thought. Another paternal aunt is known to have had breast cancer in her 40s and her paternal grandmother 'died young.' The woman is not unduly concerned about her family history because the cancers occurred on her father's side of the family. She has two sisters and a young daughter.

Understanding when genetic testing is applicable   A woman with a strong family history of colon and endometrial cancer has read about genetic testing in a women's magazine. She has decided that she wants testing for 'the cancer gene.' She has no living relatives affected with cancer.

Specific areas of genetic knowledge necessary for competent practice in cancer settings

To fulfill the competencies related to genetics in a cancer care setting, the practitioner should understand the underlying theory of genetic mutations that are associated with cancer predisposition, their implications and the current limitations of genetic testing. He/she should also have a keen awareness of the significance of family history in terms of age at cancer diagnosis, presence of multiple primary tumors and possible associated cancers. A knowledge of appropriate guidelines for client referral to specialist genetics services is essential, as is an awareness of the psychosocial impact of cancer family history, such as how families cope with living at increased risk and how 'family myths' may interfere with the processing of genetic information.

Pediatric care

Role of nurses in pediatric care

Pediatric nurses may work in neonatal or pediatric hospital wards, in specialist units such as cancer or metabolic units, or in a community pediatric setting. The nurse gives child-centered care, whilst maintaining and supporting family involvement. Increasingly, care is supported in the family home so that many pediatric nurses will have roles working in both hospital and in the community – for example, with children with cystic fibrosis or cancer.

The pediatric nurse:

- cares for children with acute illness, or whose chronic illness is exacerbated;
- supports, educates and empowers parents and the wider family in caring for the sick child;
- provides therapeutic care and support of the family unit;
- provides care for the child with special needs.

Although the main focus of the pediatric nurse is on children of all ages, care is often a partnership approach with parents so that expertise with adults in a caring role is needed. Some involvement will be short term for acute illness but, in a number of instances – particularly with chronic disability that may be genetic in origin – the nurse is likely to have longer-term therapeutic relationships with children and their families. Such situations require many psychosocial adjustments on the part of the child, the parents and close family, and the pediatric nurse is often integral in facilitating that adjustment as well as acting as an advocate for the child.

Case examples

To be effective, the general pediatric nurse will need a breadth of current genetic knowledge; in some specialist areas, such as cystic fibrosis, this will need to be of greater depth.

A genetic knowledge base is essential:

To provide supportive, informed communication at the time of diagnosis or receipt of definitive results An infant of 3 weeks is diagnosed with phenylketonuria. His parents have no experience or knowledge of the disease, and are distressed at the thought that their child is affected by a lifelong condition.

To act as a trusted and informed carer for children growing up with genetic conditions Jacob is a young man, 15 years of age. He has Duchenne muscular dystrophy and, although he lives at home, he is admitted to the pediatric unit regularly for intensive physiotherapy and management of his recurrent chest infections. Jacob feels he cannot ask his parents about his prognosis, and confides in his nurse, whom he has known for several years.

To help parents deal with feelings of anger, guilt, or blame as they acknowledge their own genetic contribution to the child's illness or consider prenatal diagnosis in a future pregnancy Marcos is a boy who is 6 years old and has hemophilia. Although his treatment is effective, the strain of the continual need for hospital appointments and the occasional admission to the pediatric ward intrudes on his life and the lives of his parents. His condition is due to an X-linked recessive gene mutation, and his mother cannot throw off the feeling that his illness is her fault, as she is the carrier. She finds it hard to talk about this but the pediatric nurse finds her in tears at Marcos' bedside at night.

To enhance awareness of conditions that may mimic non-accidental injury or abuse, such as osteogenesis imperfecta, or conditions that result in failure to thrive Susan is the daughter of Neal and Gill. At 6 months, she is only 4 kg in weight, and is very pale. Her community nurse is very concerned and wonders if Gill has adequate mothering skills and is feeding her baby properly. Susan is admitted to hospital for observation, and is found to have an unbalanced chromosome translocation.

To facilitate referral to local genetics services when appropriate During conversation about Susan's condition, the pediatric nurse becomes aware that Neal and Gill would like to have another child but are afraid of having one with a similar problem to Susan. He refers the family to a genetics counselor for investigation as to whether either of them carry a balanced translocation and, if so, to offer advice on prenatal testing.

Specific areas of genetic knowledge necessary for competent practice in pediatric settings

To work in a pediatric setting, the practitioner needs to have an understanding of the main mechanisms of inheritance and the concepts of penetrance and variable expressivity, with the ability to explain these simply and accurately to parents. This should include an understanding of common chromosomal disorders, such as trisomy 21, and the genetic basis, potential signs and symptoms, and complications of the more common genetic conditions that affect children, such as cystic fibrosis and neurofibromatosis. An awareness of the potential genetic causes of developmental delay and ability to observe and document unusual physical features that may indicate a child has a genetic syndrome are also required for competent practice.

Midwifery and obstetrics

The role of the midwife Midwives provide care during pregnancy and the perinatal period for women at home, in the community or in hospital. Midwives may be either hospital- or community-based, but may work across both environments to provide better continuity of care.

Midwives (or nurse/midwives) are specially trained to care for mothers and babies throughout normal pregnancy, during labor and in the post-partum period.

Midwives or obstetric nurses may also work as part of the team in fetal medicine units. They usually work with a physician who has a particular interest in fetal medicine, offering a variety of services. These may include detailed scanning, pre-natal diagnosis and close monitoring of 'high–risk' pregnancies.

Clinical applications of genetic knowledge

Midwives and obstetric nurses routinely counsel couples about pre-natal screening procedures that could detect a fetus affected by, or at high risk of, a genetic condition. These include:

- maternal serum screening;
- nuchal translucency scanning;
- ultrasound scanning for fetal anomaly;
- amniocentesis (for example, for advanced maternal age).

Midwives are regularly faced with conditions in either the mother or the fetus/baby that may be genetic in origin. When taking a history, the midwife needs to be alert for clues that may indicate an increased genetic risk in the pregnancy, such as:

- a history of learning difficulties in the family;
- a history of a genetic condition, such as Duchenne muscular dystrophy;
- parental ethnic background that may indicate screening should be offered for certain genetic diseases, such as Tay–Sachs disease, thalassemia and sickle-cell disease (SCD).

In addition, the mother may require special care during the pregnancy because she is affected by a relevant genetic condition, such as Marfan syndrome.

Specific areas of genetic knowledge necessary for competent practice in a midwifery or obstetric setting

Basic competence in understanding of antenatal screening procedures and their benefits, risks and limitations is a requirement for all midwives, who have the responsibility to inform and advise women about their options during the pregnancy. A knowledge of the effects of common chromosomal disorders, such as trisomy 21, must underpin discussion about screening for such conditions.

It is also essential for midwives to understand the realities of genetic testing, including the benefits and limitations. Knowledge of the genetic basis, effects and possible complications of genetic conditions that will influence care of the mother during and after pregnancy are required for effective and safe care of the mother.

Counselors for hematological conditions (hemoglobinopathies and inherited bleeding disorders)

The role of the hemoglobinopathy counselor

Hemoglobinopathy counselors care for families and individuals at risk of or affected by hemoglobinopathy by providing genetic screening and counseling for populations with and 'at risk' of hemoglobinopathies and related conditions (e.g. G6PD), community nursing support to pediatric and adult patients and their families, and specialist educational support to 'at risk' communities, health and allied professionals, non-professionals and the general population. Hemoglobinopathy counselors work with clients of all ages who have or are potentially 'at risk' of a hemoglobinopathy. This includes pregnant women and their partners, pediatric and adult patients, and the general public in low- or high-risk groups (tested and untested). They work in a variety of settings, including within specialist stand-alone units, as part of a hospital hematology or community genetics unit, and within primary care.

Clinical applications of genetic knowledge in hematological settings

General *ad hoc* testing programme   Jackson is a 23-year-old who has SCD. He is moving into a home with his girlfriend, Josie, who knows all about SCD because her cousin has the condition. Josie tells the nurse in the sickle-cell clinic that they are not using any contraception because they hope to have a family as soon as possible. The hemophilia nurse offers to refer Josie for discussion of carrier testing for herself.

Antenatal screening and counseling services   Josie is found to be a carrier, but she and Jackson discuss prenatal testing with the counselor and decide that they could not terminate a pregnancy. They therefore decide not to have prenatal testing in any future pregnancy.

Neonatal screening program   Jackson and Josie are expecting a baby. They have decided against any prenatal testing, but want to know about the screening tests for SCD in their area.

Outpatient pediatric and adult clinic   Julie's father has von Willebrand's disease, causing a disorder of his clotting mechanism. She accompanies him to the outpatient hematology clinic, and while waiting for the doctor she asks the nurse why she does not have the disease and if her children could inherit it.

Areas of genetic knowledge necessary for competent practice

Specific knowledge of the nature and variety of genetic mutations and the inheritance pattern in either hemoglobinopathies or inherited bleeding disorders is required by practitioners working in this area of healthcare. The ethical and scientific basis for conducting family studies and the importance of pre- and post-test counseling should also be understood. This will include confidentiality of genetic information, notification of genetic results, accessing, storage, and retrieval of genetic information, and the mechanisms for supporting clients.

Answers/possible responses to study questions

Chapter 2

1. 45, X — Turner syndrome
   47, XXX — Triple X syndrome
   47, XXY — Klinefelter syndrome
   47, XYY
2. (a) 46,XX
   (b) 47, XY, +8
   (c) 45, XX, (rob)13,14
3. Deletion, missense mutation, nonsense mutation, insertion.
4. Dependent on the disease selected.

Chapter 3

Task A

(a) Depigmented skin patches; Ungual fibromas; Shagreen patches; Adenoma sebaceum; Seizures; Mental retardation or learning difficulties; Cardiac rhabdomyoma; Renal cysts; Renal angiomyolipoma; Retinal hamartoma.

(b) Patients with tuberous sclerosis have calcified patches in some areas of the brain, which are detectable by CT scan. This is due to the accumulation of calcium in the subependymal nodules, which may occur in infancy.

Task B

(a) Leroy's parents:
   (i) could have been very protective of Zaina, not wishing her to be disturbed by a process she did not understand;
   (ii) may feel guilty;

(iii) may be concerned that they might be blamed for Zaina's condition;

(iv) may be worried that Leroy will decide not to have a family if the results are adverse.

The legal position varies greatly, but in the UK, it is not legal to take a sample from an adult without their informed consent. No other person can give that consent on behalf of the person. However, if the test is necessary for the person's best interests (e.g. for their medical care), this is a valid legal argument. In this case, it could be argued that there was no direct benefit to Zaina, and therefore taking a sample was not legally possible.

Chapter 4

(a) A basic ethical principle involves not acting outside the realm of your own expertise. If Janice felt she was not sufficiently skilled to 'hold' Dean's emotions safely, she acted appropriately in not encouraging him to 'open up' any further. However, as genetic issues often have an impact on the psychological health of a client, it could be argued that to act ethically in this setting a practitioner should be competent in holding the client's emotions. In this scenario, the best interest of the client would have been served by sensitively offering to refer him to an appropriate counselor or psychologist.

(b) In discussing his worries about his relationship, Dean has expressed trust in Janice. In appearing to disregard his main concern, she may have made it more difficult for him to discuss his concerns with other helpers. In this instance, it may have been preferable for Janice to openly acknowledge that she felt his concerns were outside her remit, but offer to make a referral to an appropriate practitioner who could offer him support.

(c) This response will depend on your own professional code of ethics, but would usually include listening to the client's concerns, ensuring the psychological safety of the client during the session, referring the client to an appropriate person or facility, seeking counseling and clinical supervision and undertaking further counseling training if appropriate.

Chapter 5

1.

Task A

The losses could include:

- loss of pregnancies;
- loss of potential children;
- loss of opportunity for parenthood;
- loss of reproductive confidence;
- loss of self-esteem;
- loss of self-image as a man or woman.

Task B

Responses might include:

- Eva – relief that she was not to 'blame' for the pregnancy losses; potential relief that the reason for the losses was known; potential hope that a normal pregnancy was possible.
- Juan – feelings of guilt or self-blame, feelings of inadequacy as a husband, distress at feeling powerless about the situation.

Task C

Couples who have previously experienced pregnancy loss often feel the need for additional support from a practitioner who understands the background and current risks. Offering emotional support in person or by telephone on a regular basis during the first half of the pregnancy may be helpful. Otherwise, being available for support when the couple feel they would benefit from contact is essential. Offering to see the partners separately at times may enable each to voice fears and emotions that they are unwilling to express in front of the other partner, for fear of hurting that person.

Task D

When a practitioner personally experiences a life-changing event, it may influence the nature of the counseling relationship with the client. This is especially relevant during periods of stress or loss. The practitioner often has difficulty in separating the client's emotions and experience from his or her own. This can result in a lack of 'holding' for the client. In these circumstances, it is wise to discuss the situation with a counseling or clinical supervisor and perhaps withdraw from this type of work until able to work safely (after some resolution of personal feelings).

2.

Task A

Potential second-order changes:
- For Milly: seeing herself as being vulnerable to cancer.
- For Karen: having to change her view of Milly to acknowledge her potential health issues; seeing Milly as a person who is vulnerable; perhaps seeing Milly as a person who might become dependent on her through ill-health.

Task B

You might address Milly's refusal by:
- Establishing trust through genuineness.
- Using empathic listening to hear her story and gain an understanding of her perspective.
- Gentle challenge of her beliefs about her risk (based on physical resemblance).

Chapter 6

Task A
- Think about what the ward nurses might want to know and what you think would be useful for them to know.
- If possible, contact the nurse who requested the seminar to find out what they are expecting from you. For example: signs and symptoms of Hunter syndrome, nursing care needs, possible treatments and interventions, effects on the family, genetic implications for the family.
- Before the session, find out how many nurses will be attending, where you will be giving the teaching session, what type of setting it will be and how long you have to give the talk, as these factors will determine the style of the session.
- Prepare any teaching aids, such as pictures or handouts, and think about producing a lesson plan indicating what the objectives of the session are and how you will meet those objectives.

Task B

Useful sites:
- Contact a family: http://www.cafamily.org.uk/home.html
- Gene cards: http://bioinfo.weizmann.ac.il/cards/index.shtml

Hunter syndrome (or MPS-II) is an X-linked inborn error of metabolism. It is one of a group of lysosomal storage disorders. Most children with Hunter syndrome have a severe form, with early physical signs and symptoms – including skeletal deformities, enlarged liver and spleen, and progressive heart and lung failure. They also show neurological damage that presents as developmental delay and hyperactivity but progresses. Life expectancy is normally less than 15 years of age, usually as a result of heart or lung disease. Some people with a mild form of MPS-II may survive into adulthood, with less physical signs and often with normal intellect. Currently, treatment is supportive; however, enzyme replacement therapy is under development.

Task C
Learning objectives might include:
- to understand the signs, symptoms, and prognosis of Hunter syndrome;
- to be aware of the nursing needs of the affected child and to be able to plan developmentally appropriate care;
- to be aware of the needs of the family who are caring for a child with progressive physical and mental deterioration;
- to understand the genetic basis of the disorder, identify those at risk, be aware of the psychosocial effects on the family and know where to make appropriate referrals.

Task D
The objectives for the family should include:
- establishing what the family want to know and how much they know already;
- establishing the family's expectations of the genetic counseling session;
- identifying possible 'at risk' family members;
- encouraging Pippa to access up-to-date knowledge about Hunter syndrome and also to listen to the family and be aware of their experiences, needs and expectations.

Chapter 7

Task A

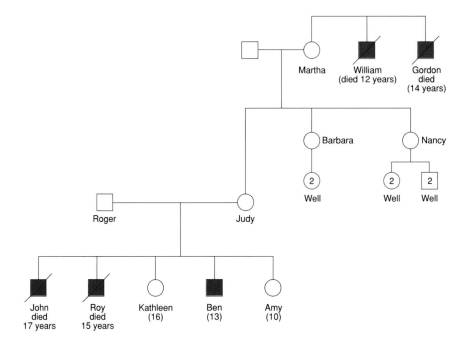

Task B

(a) Autonomy refers to self determination and the individual's right to make decisions. In this situation, Judy is acting on behalf of Ben, as he is legally a minor who cannot give consent for participation in research. Judy, or Roger, has the right to consider Ben's participation in research without fear of loss of services or loss of interpersonal interactions with Ben's physician and healthcare team. Ben also should have the opportunity to express his wishes regarding participation in research.

Non-malificence refers to the duty to do no harm. In this situation, the genetics professional has a duty to avoid harm by assuring that Judy and Ben have the opportunity to understand the purposes of the research and receive assurance that their healthcare will not be affected by their decision.

Beneficence means to do good. In many situations, families will benefit from time to consider the potential benefits and harms that may result from participation in research. Accessibility to Judy, to Ben and to others in the family who may participate in the discussions may assist Judy or Roger in determining which decision would be the best for them.

Justice refers to being treated fairly. This would only be an issue for certain patients who were offered the opportunity to participate in research based on specific characteristics – for example, based on particular socioeconomic level or on residing in an institution.

(b) If Judy could select which group she is in, she would have violated one of the components of quantitative research, which is random assignment to groups. In order for the researcher to maintain rigor of the research design, random assignment is an essential strategy, so that people with specific factors that may influence the outcome of the study, such as education, cognitive ability or emotional status, have an equal chance to be in either group.

The ethical principle of justice could also apply in this situation. Each person in the study has an equal opportunity to be assigned to the treatment or to the control group.

In discussing Judy's concerns with her, the genetics professional may ask her to describe why she would prefer the brochure-only group. As a part of this discussion, the genetics professional can explore Judy's concerns, and clarify that participation in the research is voluntary. Other opportunities for information are also available to Judy. In addition, the genetic healthcare provider can review the reasons for random assignment that are part of the basis for the research study, and to emphasize that this contributes to the ability to generalize the results. The genetic healthcare professional may also reinforce statements from the consent document that note that individuals participating in this research may, or may not, receive a personal benefit from the participation.

(c) As a member of the research team, you may discuss this situation with the person who is the principal investigator of the project. You can refer to published guidelines for the establishment of authorship of reports from research projects. You could also determine what is the traditional pattern for this decision in your own professional specialty, as well as in your institution.

Chapter 8

Task A

(a) Goals would include:
- indentifying his expectations for the genetic evaluation;
- his reasons for seeking testing;
- obtaining family history information;
- collect information on the health of the family over three generations;
- signs or symptoms that are found in HD.

(b) Questions to ask him:
- Mr. Wood's understanding of HD;
- his expectations for the genetic counselling visit;
- his expectations for genetic testing;

(c) Information to provide to him:
- information about the appointment;
- the desirability of bringing a support person to the appointment.

Task B

(a)

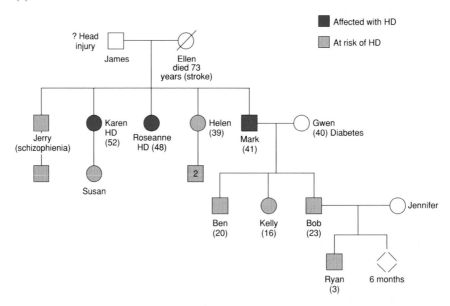

(b) Family members who may have or had HD – James, Mark, Karen, Roseanne, Jerry.

(c) Features that are consistent with autosomal dominant inheritance are HD in parent and offspring and HD in both genders.

(d) Family members who would have a 50% risk of developing HD if these people had HD – Helen, Susan, Bob, Ben, Kelly, Jerry's son.

(e) Family members who would have a 25% risk of developing HD – Helen's sons, Ryan, Bob and Jennifer's fetus.

Task C

Issues to be considered:

- autonomy of decision for Ben;
- confidentiality of medical information for Ben and for Bob;
- expectations and burden on Jennifer as a support person for two siblings;
- concern for Jennifer's emotional status and implications of results for her children.

Task D

Resources:

- HD testing protocol;
- Gene Clinics (laboratory information);
- current literature (for interpretation of test results).

Task E

Medical confidentiality:

- USA – HIPAA regulations;
- UK – Consent and Confidentiality Guidelines published by Joint Committee on Medical Genetics (JCMG);
- security of genetics clinic records.

Task F

This information would be for both brothers:

- The decision to have a predicitive test for HD is a personal decision:
- each person will have their own reasons for deciding if they want to have the test;
- the test can be done at anytime in an adult. The decision can be made now or at a time in the future;
- each person who has the test will need to have a support person with them. It is recommended that this person not be a close relative who is also at risk of developing HD. We will not divulge information about the decision to have the test, or the test results to anyone but yourself;
- we will support whatever decision you make.

Task G

(a) Bob's test results are in the abnormal range. This means he will develop HD if he has a normal life span.

(b) Ben's results are in the normal range. This means he is highly unlikely to develop HD.

(c) Each would be counseled with the person they chose as their support person. The results of the test would be explained to each of them. Each would be offered counseling to help them make the emotional adjustment to their own test result. Bob would also be informed about the availability of ongoing health monitoring for himself, according to clinical protocols. Depending upon Bob's interests at this time, further information regarding HD, support services, and healthcare could be discussed.

Chapter 9

Task A

(a) Ideally, you need to confirm the diagnosis in Alan's brother and particularly check for his genotype. This will require consent from Alan's brother.

(b) The mode of inheritance is autosomal recessive.

(c) Progressive iron overload, which may lead to liver damage, diabetes, cardiomyopathy, hypogonadism, and arthritis. Early symptoms are non-specific and include fatigue, joint pains, and loss of libido.

(d) Alan's risk of inheriting the genetic predisposition is 1 in 4. The penetrance of clinical disease is uncertain, and there is effective treatment in the form of venesection. Alan's risk of developing clinical complications is therefore considerably less. Possession of the gene mutation in the homozygous or compound heterozygous form is a major risk factor for hemochromatosis. However, hemochromatosis should be conceptualized as a clinical condition of progressive iron overload. Diagnosis is dependent on evidence of iron overload.

(e) Testing strategies will vary according to local practice. If Alan's brother is homozygous for the common mutation, then a genetic test will determine if Alan is at risk. However, follow-up iron tests – specifically, transferrin saturation and ferritin – are required for the evaluation of iron overload. Some experts recommend genetic and iron testing, whereas others would suggest only evaluation of iron status. If genetic testing is not performed, the iron tests will need to be repeated at intervals throughout Alan's life. Testing should be offered as there is an effective treatment.

Task B

(a) The early symptoms are non-specific and common in the general population; therefore, Alan cannot be certain that he has haemochromatosis.

(b) If Alan has hemochromatosis or has the genetic mutations, then his children should be offered testing when they are adults. Some families prefer that the unaffected parent has a genetic test on the assumption that the children only need to be tested if this parent carries the common mutation. This also applies to the children of Alan's brother.

(c) If Alan's test came back positive then the counselor should arrange for assessment of Alan's iron status and specialist referral. The counselor could discuss again with Alan the fact that there is effective treatment if the condition is diagnosed early enough. Alan's children should also be offered testing.

Task C

The decision to offer testing to relatives of carriers is controversial and will require discussion with the family and local health-service providers. It is not considered appropriate to offer population screening for this condition; therefore, careful consideration should be given to the appropriateness of actively offering testing outside of the immediate family.

Chapter 10

Task A

This is not reassuring, as it would be possible for Michelle to be a carrier and have normal clotting factors. Michelle could have a 50% chance of being a carrier, dependent on whether her brother is a new mutation or not. If she is a carrier, any female child will have a 50% chance of being a carrier, and any male child will have a 50% chance of being affected. Her risk of having an affected child ranges from very low (if her brother has a new mutation) to 25%.

Task B

The testing that is available will depend on local facilities. If a mutation has been identified in her brother, Michelle may be able to have direct mutation analysis for herself or as a prenatal diagnosis. If mutation testing is not available, linkage analysis may help determine her carrier status.

Michelle's options include:
- having no testing and therefore taking the risk that her baby will have hemophilia;
- establishing her carrier status and then deciding about prenatal diagnosis;
- having prenatal testing either with mutation testing (if available) or with fetal blood testing if the baby is male.

Task C

As Michelle is already pregnant, it is important to establish what her feelings are about whether she wants any further testing at all. Her perception of the severity of the condition will be influenced by the additional complication of her brother's HIV status. This should be explored in the counseling. In this type of genetic counseling situation, where there is a time pressure because of the need to consider prenatal diagnosis, it is important for the counselor to be aware of the need to be non-directive and elicit, and listen to the client's concerns. The above are suggestions as to what issues may be most important; there will be others to consider.

Chapter 11

1.

(a) Yes, bilateral breast cancer in a first-degree relative under the age of 40 years would put Lynne into the high-risk group.

(b) No, his parents are not biologically related and the two cancers would not significantly increase his risk to the level at which additional screening would be suggested. His risk of colon cancer would be about 1 in 17.

(c) No, it is unlikely that the cases of cancer are due to a familial mutation, especially as the lung cancer was diagnosed in a heavy smoker.

(d) Yes, this history may be indicative of an HNPCC mutation in the family.

(e) Yes, breast, ovarian and pancreatic cancer are known to occur in individuals with a *BRCA2* mutation, and Barry's sister was affected at the young age of 39 years.

2.

Task A

A full four-generation family tree, and the confirmation of the diagnosis and the age of diagnosis in all affected relatives.

Task B

Microsatellite instability studies and DNA analysis can often be performed on tissue stored from deceased affected relatives, such as surgical specimens. If the mutation is detected, presymptomatic testing can be offered to James.

Task C

The response depends on current guidelines for screening in your own region, but 2-yearly colonoscopy would usually be appropriate from the age of 25 years.

Chapter 12

1.

Task A

There is an agreed testing strategy for Angelman syndrome and Prader–Willi syndrome, which includes tests for methylation status, uniparental disomy, imprinting gene mutations and karyotyping. Exactly what tests are available will depend on local facilities.

Task B

A balanced rearrangement of chromosome 15 could lead to deletion of the critical region in the unbalanced form. Inheritance of the derived chromosome from the mother would lead to Angelman syndrome and from the father would result in Prader–Willi syndrome.

2.

(a) You would want to explore her feelings about her brothers diagnosis and if she would regard this as an appropriate reason for prenatal diagnosis and possible termination of a pregnancy. Her previous knowledge and expectations would also need to be explored.

(b) Ideally, the karyotype of her brother.

(c) If her brother has regular trisomy 21, her risk is no higher than that of the general population. She could choose to have no testing, have some kind of screening procedure or proceed directly to amniocentesis or chorion villus biopsy.

Chapter 13

1.

Task A

A family tree documenting the history of dementia, including age of onset and characteristics (e.g. any neurological abnormalities). If possible, medical records of affected individuals should be obtained to establish if a specific diagnosis has been made.

Task B

The features of a high-risk family tree are multiple affected members, early age of onset, and a clear Mendelian inheritance pattern (autosomal dominant).

Task C

If the family tree indicates autosomal dominant inheritance then a predicitive testing protocol should be followed if testing is possible. In both situations the counselor might want to explore with Martin his anxieties and worries about developing Alzheimers disease, and establish with him a plan for developing strategies to cope with those.

Web-based resources for genetic healthcare professionals

These resources were checked before publication, but are obviously subject to rapid change. Please check these resources before suggesting them to clients.

This list covers a range of general sites, plus some disease-specific sites. Other website resources are listed at the end of each relevant chapter.

Title	*Website*
General sites	
American Academy of Pediatrics. Search engine, plus research information.	http://www.aap.org/
Cancer and Genetics. General information on cancer.	http://www.cancergenetics.org/ genetics.htm
Center for the Study of Autism. Information on all syndromes with autistic tendencies.	http://www.autism.org/
Familial Cancer Syndromes a Focus of Cancer Genetics Research. General information site.	http://audumla.mdacc.tmc.edu/ ~oncolog/09_Familial.html
Gene Clinics/Gene Reviews. Information on many genetic conditions and updated information on genetic testing.	http://www.geneclinics.org
Genetic and Rare Conditions Site. Information on lots of disorders.	http://www.kumc.edu/gec/support/
Genetic Alliance. Information on genetic support groups and links to information on range of conditions.	http://www.geneticalliance.org

Karolinska Institutet, Sweden.   http://info.ki.se/index_en.html
Search engine for latest research.

Medical search engine.   http://www.mdchoice.com/pt/
                                                 index.asp

National Disease Information Center.   http://www.diseases.nu/index.htm
Information on physical diseases and
mental disorders.

National Institutes of Health.   http://www.nih.gov/
Scientific resources.

National Library of Medicine.   http://www.nlm.nih.gov/
For research and use of Medline.

National Organization for Rare   http://www.rarediseases.org/
Disorders.
Database.

Parent Matching and Support Groups.   http://www.nas.com/downsyn/
Parental and Sibling Support.   parent.html
Addresses for several conditions.

Professional organizations
(Please also see Chapter 1, Table 1.1 for a more complete list of professional organizations)

American Society of Human Genetics.   http://www.ashg.org

International Society of Nurses in   http://www.isong.org/
Genetics.

British Society for Human Genetics.   http://www.bshg.org

American College of Medical Genetics.   http://www.acmg.net

Association of Genetic Nurses and   http://www.agnc.org
Counsellors.

National Society of Genetic Counselors.   http://www.nsgc.org

Adult polycystic kidney disease

Polycystic kidney disease.   http://www.kidney.ca/poly-e.htm
Very clear and concise information.

Polycystic Kidney Disease.   http://www.mdchoice.com./pt/
People-friendly information.   ptinfo/pkid.asp

Polycystic Kidney Research Foundation.   http://www.pkdcure.org/
Lots of general information, links,
support groups, news and research.

Vanderbilt Medical Center.   http://www.mc.vanderbilt.edu/
Lots of technical information.   peds/pidl/rephro/polykidn.htm

Angelman syndrome

Angelman Syndrome.
Information for families and professionals.

http://asclepius.com/angel/

Angels Among Us.
What it's like to have an 'angel' in your
family? Addresses, information and links.

http://shell.idt.net/~julhyman/
angel.htm

Facts About Angelman Syndrome.
Information, links and references.

http://www.asclepius.com/angel/
asfinfo.html

Cleft lip and palate

Cleft Lip and Palate Association.
Comprehensive guide to information,
medical care and links.

http://www.clapa.mcmail.com/

A cleft lip and palate support group.
Support group, with friendly information.

http://www.cleft.org/Smiles

Widesmiles.
Cleft lip and palate resource.
Comprehensive guide to information,
medical care and links.

http://www.widesmiles.org/
Default.htm

Cystic fibrosis

Cystic Fibrosis Foundation.
Latest news, research and information.

http://www.cff.org/

Cystic Fibrosis.

http://vmsb.csd.mu.edu/
~5418lukasr/cystic.html#C

Cystic Fibrosis.
Medical information, plus good
children's page.

http://cysticfibrosis.com/

The Creon game!
Game to download for children to
encourage them to take their enzyme
capsules with food.

http://www.cysticfibrosis.co.uk/
game.htm

Down syndrome

National Association for Down Syndrome
(NADS).
Counseling and support service for
parents of children with Down
syndrome.

http://www.nads.org/

National Down Syndrome Society.　　　　http://www.ndss.org/
A comprehensive, online information
source about Down syndrome.

Recommended Down Syndrome Sites　　　http://www.ds-health.com/
on the Internet.　　　　　　　　　　　ds_sites.htm
Quick access to lots of sites.

Welcoming Babies with Down syndrome.　http://www.nas.com/downsyn/
Friendly site, nice opening message.　　welcome.html

Familial adenomatous polyposis (FAP)

FAP. A Guide For Families.　　　　　　http://www.mtsinai.on.ca/
Information on all aspects of FAP.　　　familialgican/FAPEnglish/fap.html

Gene clinics: FAP.　　　　　　　　　　http://www.geneclinics.org/
Lots of factual information; fairly　　　profiles/fap/
readable.

Genetic Testing and Counseling in　　　http://intouch.cancernetwork.com/
Familial Adenomatous Polyposis.　　　　journals/oncology/o9601e.htm
Online paper; concise information.

Scientific Articles: Familial Adenomatous　http://www.ncgr.org/gpi/odyssey/
Polyposis.　　　　　　　　　　　　　　colon/fap_arts.html
Useful articles.

Familial breast/ovarian cancer

Familial Breast Cancer Publications.　　http://norp5424b.hsc.usc.edu/
Links to recent articles.　　　　　　　fbcpubs.html

Family Cancer and Genetic Testing.　　http://www.familycancer.org/
Information, links and support groups　FamilyCancer/index_gene.stm
for breast and ovarian cancer.

Gilda Radner Familial Ovarian Cancer　http://rpci.med.buffalo.edu/
Registry.　　　　　　　　　　　　　　departments/gynonc/grwp.html
Comprehensive information.

Ovarian cancer.　　　　　　　　　　　http://www.gyncancer.com/
Technical and basic information.　　　　ovarian-cancer.html

Hereditary motor sensory neuropathies (HMSN)

Charcot Marie Tooth International UK.　http://www.cmt.org.uk/
Support group, with basic information.

Information on research and treatment.　http://www.ultranet.com/~smith/
　　　　　　　　　　　　　　　　　　CMTnet.html CMTnet.

Hereditary Motor Sensory Neuropathies, Charcot Marie Tooth.
Range of disorders covered; technical information.

http://www.neuro.wustl.edu/neuromuscular/time/hmsn.html

Muscular Dystrophy Campaign.
People-friendly information on HMSN.

http://www.muscular-dystrophy.org.uk/information/Key%20facts/hmsn.html

Hereditary non-polyposis colorectal cancer (HNPCC)

Hereditary Non-Polyposis Colon Cancer.
Questions and answers on a case study.

http://www.cancergenetics.org/hnpcc.htm

M.D. Anderson Cancer Center.
Hereditary colon cancer.
Basic information.

http://www.mdacc.tmc.edu/~hcc/

Penrose Cancer Center.
Hereditary Nonpolyposis Colon Cancer (HNPCC) questions and answers.
People-friendly language.

http://www.penrosecancercenter.org/info-hnpcc.htm

Scientific Articles: Hereditary Nonpolyposis Colorectal Cancer.
Useful articles.

http://www.ncgr.org/gpi/odyssey/colon/hnpcc_at.html

Huntington disease

Caring for People with Huntington Disease.
Information and links.

http://www.kumc.edu/hospital/huntingtons/

Facing Huntington Disease.
Useful information.
harvard.edu/mcenemy/facinghd.html

http://www.neuro-chief-e.mgh.

Huntington's Disease Advocacy Center.
Articles, stories, chatroom and links.

http://www.hdac.org/

UK's Huntington Disease Association.
Information, news and events.

http://www.hda.org.uk/

Hypertrophic cardiomyopathy

Hypertrophic Cardiomyopathy Association.

http://www.hcma-heart.com/

Hypertrophic Cardiomyopathy Association.
Symptoms, treatment and information.

http://www.kanter.com/hcm/

The Management of Hypertrophic
Cardiomyopathy. Review Article.
Online article.

http://www.gilead.org.il/hcm/

Neural tube defects

Folate, a guide for health professionals.
Information biased towards Australian
population.

http://hna.ffh.vic.gov.au/phd/
folate/hlthprof.htm

Maryland Dietetic Association.
Factsheet on folic acid.

http://www.eatwellmd.org/
folic.htm

Prevention of Neural Tube Defects.
Comprehensive information on neural
tube defects.

http://thearc.org/faqs/folicqa.html

Screening for Neural Tube Defects –
Including Folic Acid/Folate Prophylaxis.
Up-to-date information.

http://cpmcnet.columbia.edu/
texts/gcps/gcps0052.html

Neurofibromatosis (NF)

American Academy of Pediatrics.
Committee on Genetics.
Health supervision for children with NF.

http://www.aap.org/policy/
00923.html

Neurofibromatosis Resources.
Site set up by person with NF2,
information on NF, plus links.

http://www.neurosurgery.mgh.
harvard.edu/NFR/

Neurofibromatosis, Inc.
Information, support and links.

http://www.nfinc.org/

The National Neurofibromatosis
Foundation, Inc.
Information, links and support groups.

http://www.nf.org/

Prader–Willi syndrome

Healthcare Guidelines for Individuals
with Prader–Willi Syndrome.
Recommendations and references.

http://www.pwsausa.org/postion/
HCGuide/HCG.htm

Prader–Willi Syndrome.

Full run-down of clinical features,
plus history of disorder.

http://www.icondata.com/health/
pedbase/files/PRADER-W.HTM

Prader–Willi Syndrome.
People-friendly information on disorder.

http://www.geneclinics.org/
profiles/pws/

The Prader–Willi Syndrome Association (USA).
Information, links and support groups.

http://www.pwsausa.org/

Smith–Lemli–Opitz syndrome

Smith–Lemli–Opitz Syndrome.
Factual information.

http://www.med.jhu.edu/CMSL/SLOS.html

Smith–Lemli–Opitz Syndrome.
People-friendly information on syndrome.

http://www.geneclinics.org/profiles/slo/

Smith–Lemli–Opitz/RSH syndrome.
Information on clinical aspects.

http://members.aol.com/slo97/

Smith–Magenis syndrome

Smith–Magenis Syndrome.
Lots of information, links, references and support groups.

http://www.kumc.edu/gec/support/smith-ma.html

Smith–Magenis Syndrome.
Quick run-down of clinical features.

http://www.bcm.tmc.edu/neurol/research/genes/genes10.html

Smith–Magenis syndrome.
Contact a family site. Information and support.

http://www.cafamily.org.uk/Direct/s33.html

Spinal muscular atrophy

Families of Spinal Muscular Atrophy.
Information, links and support.

http://www.fsma.org/

Families of Spinal Muscular Atrophy.
Quick guide to spinal muscular atrophy.

http://www.fsma.org/booklet.htm

Spinal Muscular Atrophy.
User-friendly information.

http://www4.ccf.org/health/health-info/docs/1300/1346.HTM

Tuberous sclerosis

Tuberous Sclerosis Association.
Latest news, information, links and funding.

http://www.tuberous-sclerosis.org/

The National Tuberous Sclerosis Association.
Information, resources and services.

http://www.ntsa.org/guests/main.web

What is Tuberous Sclerosis?
People-friendly information.

http://www.insteam.com/CT_TSA/TSdef.htm

Williams syndrome

Healthlink USA, on Williams syndrome. Links to several sites.

http://www.healthlinkusa.com/335.html

Williams Syndrome Foundation UK. Information, links and support groups.

http://www.williams-syndrome.org.uk/

Williams Syndrome. Contact a family site. Information and support

http://www.cafamily.org.uk/Direct/w15.html

Chromosome 22q

Telomere. References on this subject.

http://www.genes.uchicago.edu/telomere/22q.html 22q

Chromosome 22 Disease List. List of all diseases known to map to chromosome 22, with OMIM numbers.

http://www.sanger.ac.uk/cgi-bin/c22_diseases_table.pl

Chromosome 22 Central. Parent support group for chromosome-22-related disorders. Information, links and support.

http://www.nt.net/~a815/chr22.htm

Support Group site.

http://www.ucfs.net UK 22q11

Glossary of terms

Acrocentric chromosome: A chromosome that has its centromere near to one end.

Affected individual: A person who has the signs and symptoms of the genetic condition.

Algorithm: A set of decision rules that are used to address a problem.

Allele: A copy of a gene or DNA sequence at a particular locus. One allele is inherited from each parent.

Amniocentesis: Withdrawal of amniotic fluid from the amniotic sac, usually for the purpose of testing the fetal chromosomes.

Andragogy: Theory of adult learning.

Anencephaly: Failure of the anterior neural tube to close properly during very early intrauterine life, resulting in the absence of the cerebral hemispheres and skull bone, together with a rudimentary brain stem.

Aneuploidy: An alteration in the number of chromosomes, involving only one or several chromosomes rather than the entire set of chromosomes.

Anticipation: The phenomenon whereby successive generations of a family manifest a genetic condition more seriously or at a younger age.

Assisted reproduction: Any artificial technique used to enable a pregnancy to be achieved (e.g. *in vitro* fertilization).

Autism: A form of mental disability that is characterized by failure to interact with others.

Autosomal dominant inheritance pattern: The inheritance pattern whereby one copy of a gene is mutated; this is sufficient to cause the disease to be manifested.

Autosomal recessive inheritance pattern: The inheritance pattern whereby both copies of the gene are mutated and the person develops the condition because they have no normal copy. Carriers of recessive conditions are usually unaffected.

Autosomes: The chromosomes that are present in equal numbers in both the male and female of the species (in humans, the chromosomes 1 to 22).

Base pair: A pair of nucleotides, which are positioned opposite each other on the two strands of the DNA double helix. Adenine always pairs with thymine, and guanine with cytosine.

Carrier: A person who is generally not affected with the condition, but carries one mutated copy of a gene. Generally relates to heterozygotes in recessive or X-linked conditions.

Chiasma: The point at which two homologous chromosomes cross over during meiosis.

Chorionic villus biopsy: Removal of cells from the chorionic villi (developing placental tissue).

Chromatid: One of two lengths of chromosomal material (sister chromatids) that are joined at the centromere during cell replication. Each becomes a new chromosome.

Chromosome: The physical structures into which the DNA is packaged within the nucleus of cells. The usual number of chromosomes in humans is 46.

Clinical genetics: The branch of the health service that is chiefly involved in diagnosis of genetic conditions and genetic counseling for families.

Codon: A triplet in the messenger RNA that provides the code for one amino acid.

Colectomy: A surgical operation to remove the colon.

Colonoscopy: An investigation wherein the rectum, sigmoid and large colon are viewed directly via an endoscope.

Consanguinity: The biological relationship between two individuals who have a common ancestor (e.g. cousins).

Consultand: The person seeking genetic information – not necessarily the affected person in the family, who is usually called the proband.

Cordocentesis: Removal of a sample of fetal blood from the umbilical cord during pregnancy.

Cytogenetics: The study of chromosomes, in the laboratory.

Delay – developmental or learning delay: A term used to describe the failure of the child to reach milestones in physical, mental, emotional or social development within the expected age limits.

Deletion: The omission of a part of the genetic material; the term can be used in relation to either a gene or a chromosome.

Diploid: Having two copies of each autosome.

Disjunction: The separation of the replicated copies of the chromosomes into two daughter cells during the second stage of meiosis.

DNA: Deoxyribonucleic acid. The biochemical substance that forms the genome. It carries in coded form the information that directs the growth, development and function of physical and biochemical systems. It is usually present within the cell as two strands with a double-helix confirmation (see Chapter 4).

Dominant: See autosomal dominant.

Duplication: The abnormal repetition of a sequence of genetic material within a gene or chromosome.

Dysmorphic features: Physical features that are outside the variability of the normal population. They may occur because of a change in the genetic code providing instructions for those features.

Eugenics: The study and practice of principles that will improve the genetic health and fitness of a population.

Exclusion test: A genetic test that uses samples from three generations of the family, to exclude the risk of a genetic disease or to confirm a 50% or 25% risk.

Exon: A sequence of DNA that contributes to the protein product of a gene (see also intron).

Expansion: An abnormally large repetition of specific DNA sequences within a gene.

Expression: The way in which the gene mutation is manifested within an individual.

Fibroblast: A type of immature connective-tissue cell.

FISH (fluorescent *in situ* hybridization): A technique that uses both cytogenetics and molecular biology to identify subtle changes in chromosome structure.

Gamete: A cell formed in the reproductive organs from the germline; in humans, these cells are either ova or sperm.

G-banding: A technique of staining the chromosomes to enable identification by creating a different pattern of bands along each chromosome.

Gene therapy: Therapy that is based on the principle of replacing or modifying a faulty genein the relevant tissues. The aim is to reduce or obliterate the effects of the genetic condition.

Gene: The fundamental physical and functional unit of heredity consisting of a sequence of DNA.

Genetic counselor: A person whose main professional role is to offer information and support to clients who are concerned about a condition that may have a genetic basis.

Genetic screening: This term usually refers to population screening for a genetic variation or mutation.

Genogram: A family tree that includes information about the nature of the relationships between individuals.

Genomics: The study of interactions between genetic and environmental factors that contribute to disease.

Genotype: The genetic make-up of an individual or the specific gene structure at one locus.

Guthrie test: A blood test performed in the neonatal period to detect infants at high risk of phenylketonuria. A test for congenital hypothyroidism is usually performed at the same time.

Haploid: Having one copy of each autosome.

Hemizygous: Having only one copy of a gene or sequence of DNA at a specific locus.

Hemoglobinopathy: A genetic condition that affects the structure of the hemoglobin molecule.

Hemoglobin S: The variant form of hemoglobin found in carriers or affected persons with sickle cell disease. The mutation being a change in single nucleotide (U to A) in the β-globin gene leading to a substitution of glutamic acid for valine at position six in the protein.

Heterogeneous: Consisting of dissimilar components; in this context, more than one gene.

Heterozygous: Having two different alleles at a genetic locus, usually one normal and one faulty copy of a gene (see also homozygous).

Homologous pair: Two copies of the same chromosome.

Homozygous: Having two identical alleles at a genetic locus. In Mendelian diseases, these may be copies of a gene that are either both normal or both faulty.

Hybridization: Attachment of one DNA sequence to an identical sequence. This mechanism is used to attach a DNA probe to a segment of genomic DNA.

Hydrocephalus: The presence of excessive cerebrospinal fluid in the ventricles of the brain, normally leading to enlargement of the head.

Imprinting: The phenomenon whereby the two copies of a gene have a different function, depending on their parental origin.

Induced abortion: Termination of pregnancy.

Insertion: The introduction of additional genetic material into a gene or chromosome.

Intron: A sequence of DNA that does not contribute to the code for protein product, as the genetic sequence within introns is omitted when the messenger RNA is made.

Inversion: An alteration in the sequence of genes along a particular chromosome. In a paracentric inversion, the change occurs on only one side of the centromere. In a pericentric inversion, the centromere is involved and material will move from the long to short arm and vice versa.

Karyotype: A description of the chromosome structure of an individual (assessed during metaphase), including the number of chromosomes and any variation from the normal pattern.

Linkage: The phenomenon whereby alleles that are physically close together on a chromosome will tend to be inherited together. This allows for a technique of genetic testing that tracks a specific copy of a gene through a family.

Locus: The position of a gene, a genetic marker or a DNA marker on a chromosome.

Mastectomy: A surgical procedure to remove the breast.

Maternal serum screening: A method of detecting a relative risk of Down syndrome, some other chromosomal abnormalities and neural-tube defects in a pregnancy, using by biochemical testing of the mother's blood.

Meiosis: The production of gametes (haploid cells).

Mendelian disorder or Mendelian condition: A genetic disorder that is caused by a single gene mutation, following a dominant, recessive or X-linked pattern of inheritance.

Messenger RNA: The sequence of base pairs that transfer the genetic code from the DNA to a functional protein.

Methylation: Addition of a methyl group to the DNA molecule to signal that the gene should not be expressed.

Microdeletion: A minute deletion of chromosomal material that is usually not detectable with a light microscope and has to be identified using a method such as FISH.

Mismatch repair gene: A gene that has the function of detecting and repairing errors in DNA transcription.

Mitochondrial DNA: The genetic material in the mitochondria, outside the nucleus of the cell.

Mitosis: The production of somatic diploid cells.

Molecular genetics: The study of genetic material at a molecular level, including DNA studies.

Monosomy: Having only one copy of a particular chromosome.

Mosaicism: Having more than one cell line with different chromosomes or expressing different genes.

Multifactorial: A condition is said to be multifactorial if both genetic and environmental influences are thought to contribute to its etiology.

Mutation: A gene-sequence variation that is found in less than 1% of the total population. The mutation may cause a change in the protein product of the gene, and therefore cause health problems for the person concerned.

Neonatal death: The death of a baby, who has shown signs of life, before the age of 28 days.

Neural-tube defect: An abnormality of the spinal column or cranium (spina bifida or anencephaly).

Neural tube: Nervous system structures that become the brain and spinal cord.

New genetics: A term used to denote a change in the field of genetics, in which the focus shifts from rare conditions caused by a fault in a single gene (including the 'Mendelian' conditions) to application of genetics to common diseases.

Non-directiveness: A model of counseling used in genetics counseling, which emphasizes the right of clients to make decisions without coercion from others.

Non-disjunction: Failure of the two copies of chromosomes to separate effectively into the two daughter cells.

Oopherectomy: A surgical procedure to remove the ovary.

Paracentric: Affecting genetic material on only one arm of the chromosome.

PCR or Polymerase chain reaction: A laboratory method of manufacturing many copies of a sequence of DNA.

Pedagogy: The study of teaching methods.

Pedigree: Family tree.

Penetrance: The proportion of persons in a population who have manifestations of a particular gene mutation.

Pericentric: Affecting genetic material on both arms of the chromosome, around the centromere.

Pharmacogenetics: The science of using the genetic variability in the population to target medication selection and dosage more effectively.

Phenotype: the clinical manifestation (signs and symptoms) of the condition.

Polygenic: Relating to a number of different genes (e.g. a disorder is polygenic if it could be caused by a combination of mutations in several different genes).

Polymerase chain reaction: see PCR.

Polymorphism: Normal variation in the sequence of DNA in a gene. It differs from mutation in that it is usually found in more than 1% of the population.

Polyp: A small tumor growing from the surface of mucous membrane.

Polypeptide: A molecule consisting of a number of amino acids.

Population screening: Using a test to assess the risk or presence of a disorder in an entire section of the population (e.g. neonatal screening for hypothyroidism).

Premutation: An alteration in gene structure that does not produce manifestation of disease in the individual carrying the premutation but may be passed on as full mutation to the next generation.

Proband: The affected person in the family or the person who is seeking genetic advice.

Probe: A sequence of manufactured DNA that attaches to an identical sequence in the genomic DNA for the purposes of genetic testing or research.

Prophylactic: Used as preventative measure.

Recessive: See autosomal recessive.

Reciprocal translocation: Exchange of chromosomal material between at least two chromosomes.

Recombination: The creation during meiosis of a newly constituted chromosome or sequence of DNA, which is a unique combination of the parent's maternal and paternal DNA.

Recurrence risk: The chance that a genetic condition will occur again in offspring or siblings of an affected person.

Restriction fragment length polymorphisms (RFLPs): Fragments of DNA within a gene that have a normal variability in size when cut with specific enzymes.

Ribosome: Ribonucleic-acid-containing particle situated in the cytoplasm that is involved in protein synthesis.

Robertsonian translocation: An attachment of two acrocentric chromosomes end to end at the centromere.

Salpingo-oophrectomy: A surgical procedure to remove the fallopian tube and ovary.

Single nucleotide polymorphisms (SNPs): Single base changes in DNA sequence that may be used as polymorphic markers.

Somatic: Relating to cells other than the germline.

Southern blotting: A laboratory method of DNA analysis.

Spina bifida: An interruption to the spinal column, with possible herniation of the spinal cord and meninges (myelomeningocoele). One form of neural-tube defect (another being anencephaly).

Spontaneous abortion: Loss of a pregnancy without interference; miscarriage.

Stillbirth: A fetus of more than 24 weeks gestation who is born dead.

Syndrome: A number of physical features or abnormalities that fit a recognized pattern.

Teratogen: A substance that may harm the developing fetus.

Translocation: An alteration in the usual structure of a chromosome, wherein part or all of one chromosome is attached to another.

Trinucleotide repeat: A sequence of three bases that is repeated more than once at a site within a particular gene.

Trisomy: Having three copies of a particular chromosome.

Tumor-suppressor gene: A gene that has the function of preventing the overgrowth or abnormal growth of cells.

Ultrasound scanning: Investigation of physical structures using ultrasound device (sound waves).

Uniparental disomy: Inheritance of both copies of a particular chromosome from one parent only.

Vertical transmission: Passing of the condition from parent to offspring.

VNTR polymorphisms (variable number tandem repeats): Variations in the number of repeat sequences of DNA at a specific locus that can be used as polymorphic markers.

Wild type: Standard, functional allele.

X-inactivation: In human females, the early random inactivation of one of each of the X chromosomes, allowing expression of genes only on the active X chromosome.

X-linked inheritance pattern: A pattern of inheritance whereby the mutated gene is on the X chromosome, of which males have one copy and females have two.

Index